Icons and Symmetries

Icons
and
Symmetries

Simon L. Altmann

Brasenose College, Oxford

CLARENDON PRESS · OXFORD

1992

Oxford University Press, Walton Street, Oxford OX2 6DP
Oxford New York Toronto
Delhi Bombay Calcutta Madras Karachi
Petaling Jaya Singapore Hong Kong Tokyo
Nairobi Dar es Salaam Cape Town
Melbourne Auckland
and associated companies in
Berlin Ibadan

Oxford is a trade mark of Oxford University Press

Published in the United States
by Oxford University Press, New York

A catalogue record of this book is available from the British Library

Library of Congress Cataloging in Publication Data
Altmann, Simon L., 1924–
Icons and symmetries/Simon L. Altmann.
Includes bibliographical references and index.
1. Symmetry (Physics) I. Title.
QC174.17.S9A48 1992   530—dc20   91-37489
ISBN 0–19–855599–7

Typeset by Cotswold Typesetting Ltd, Cheltenham
Printed in Great Britain
by Biddles Ltd,
Guildford & King's Lynn

Preface

In May 1989 Professor Arnout Ceulemans asked me whether I would be prepared to give three lectures on symmetry for the Faculty of Science of the Catholic University of Leuven, and he suggested as an example of what they had in mind the famous book by Weyl. My immediate reaction was that I would certainly enjoy giving such lectures but that I should not attempt to follow in the steps of the master. My view was that it was well nigh impossible to improve on what Weyl had done in that style, and that a number of books had already appeared where the very general, one might call them cultural, aspects of symmetry had been extremely well covered. I instead suggested that I would like to take three case-studies in which, by working through the history as well as through the physics of the problems in hand, one would learn how some of the major concepts of symmetry became established in our scientific culture.

I was aware, of course, that in adopting that format I would have to constrain myself to a very limited cross-section of the major ideas on symmetry, but I felt that this would be compensated by the fact that, addressing a general audience of physical scientists, one would then be able to give a reasonable account not only of some of the basic ideas but also of how they can actually be applied in order to obtain useful results. I also believe that the best way to show the relevance of philosophical and historical arguments in science is through specific case-studies, where the understanding of a physical problem, discussed at some depth, is greatly enhanced in the context of one given historical time and of the then prevailing philosophical ideas.

Because of this approach, the reader cannot expect to find here many of the most dramatic applications of symmetry, say to elementary particles or to the study of magnetic crystals. Even the word *group* will not be found in this book, although group theory is undoubtedly the most important mathematical tool in the study of symmetry. This omission, however, is deliberate, because the level at which group theory can illuminate the concept of symmetry is such that, in a general treatment such as this, one would have to fall back on the use of mere words rather than of the ideas behind them. Given all these limitations, it will be necessary for me to say a little about what the reader will actually find in this book.

I discuss in Chapter 1 the origins and applications of the principle of symmetry, illustrating it in particular with the famous Ørsted paradox of the interaction between the magnetic needle and the electric current. This will teach us to understand that vectors have very specific symmetry properties, that the nature (and the name) of a vector depends on its symmetry behaviour, and that this symmetry behaviour must be determined experimentally. A very important result of our discussion will be, in fact, that certain experimental observations must be regarded as experiments conducted in order to determine some specific symmetry properties, thus dispelling the notion that symmetry in physics belongs in the realm of the *a priori*. I also hope that the discussion of null experiments from the point of view of conventionalism will throw new light on this old problem. All this is interwoven with the wonderful story of Ørsted, and some hitherto un-published material from his manuscripts that provides a new insight into the controversial circumstances of his discovery will be found here. I hope that this will be of some interest to historians of science.

Whereas Chapter 1 is mainly concerned with reflection symmetry, Chapter 2 deals with rotations and with the way in which Hamilton tried to describe them by means of quaternions. This will give us a further insight into the symmetry properties of vectors, it will show the way in which tensors gradually moved into the picture, and finally it will explain how spinors were born. This story will contrast the algebraic treatment of rotations introduced by Hamilton with the geometrical one of Rodrigues, and we shall see how the diametrically opposed philosophical outlooks of these two authors profoundly influenced their view, and thus our view, of rotations and of vectors.

A very important idea underlying the whole book is strongly illustrated in Chapters 1 and 2, and this is the concept of an *icon*, the importance of which I believe has not been sufficiently recognized so far in philosophy of science, although Reichenbach came very near it in discussing the visualization of geometry. In order to describe physical objects, we all know that we use more or less idealized models. Very often, these models are graphically depicted by icons, such as the arrows used to denote vectors. Chapters 1 and 2 both deal, in different situations, with the subtle relation between physical objects and their icons. I argue in Chapter 1 that the notion that symmetry is a geometric *a priori* property, a notion which has worried humanity since the days of Archimedes, is largely due to the confusion between the object and its icon. In Chapter 2 I illustrate on the other hand, with the historical example of Hamilton's early definition of vectors, how it is possible to read the icon of a given object as if it were that of an entirely different one.

Chapter 3 attempts to do two jobs, the first of which is to illustrate how symmetry can be applied in order to study the ways in which energy levels are classified in atoms and solids. This chapter, in fact, contains just about the most compact treatment of bands and Brillouin zones in solids that I know of. This treatment is applied to our second job, which is to understand how symmetry, which is always conserved in the first two chapters, can now be broken. I treat for this purpose the Peierls transformation in linear and quasi-linear chains.

If this little book manages to convey the importance of the distinction between models and their icons and if it reveals to the reader how much of our scientific culture depends on the often taken for granted use of icons, then I would feel contented that one of my major jobs has been done. And all this will greatly contribute to the important task of firmly establishing symmetry as an empirical property to be experimentally determined.

I should like to say a few words about the type of reader to whom this book is addressed. Chapter 1 is largely non-mathematical and I hope that it will attract a fairly wide audience with interests ranging from philosophy and history of science to physics. Although I use some mathematics in Chapters 2 and 3, this is quite elementary and self-contained, of the type learnt at the sixth form in the United Kingdom. Indeed, I hope that this book might be read by some young people even before they come up to university, in order to widen their horizons a little after a couple of years of severe specialization. Science undergraduates in their junior years might also enjoy reading the book in their spare time, and I hope that it might be of some use in some of the more general science courses available in American universities and liberal arts colleges. I have very much in mind, also, the more mature scientist, who graduated perhaps in the early seventies when symmetry was hardly touched upon in degree courses and who would like to have some idea of what the whole thing is about.

I am most grateful to the Faculty of Science of the Catholic University of Leuven for nominating me for the Vlaamse Leergangen Chair for 1989 and to the Board of Vlaamse Leergangen for appointing me to it. I should like to thank in this respect the Dean of the Faculty, Professor A. Dupré, the Chairman of the Vlaamse Leergangen Board, Professor P. Tobbak, and Professors Arnout Ceulemans and H. Reynaers, who looked after all the practical details of my visit to Leuven with unfailing kindness. The three lectures I gave during the tenure of the Chair at Leuven in October 1989 were a most welcome challenge for me, and if readers have any fun in reading this book I hope that they will bear in mind, as I do, the wonderful audience with which I tried to share my enthusiasm and delight in these subjects. I felt that it would have been a pity to destroy the informality of the lectures and so I have kept the book very close in style to them.

It is a pleasure to acknowledge the most generous help of the Royal Library of Copenhagen. I am very grateful to Dr Tue Gad, of its Manuscripts Department, for having arranged to detach Ørsted's correction to page 16 of the manuscript of 'Thermoelectricity', and for kindly providing me with Xerox copies of the result. Mr Erik Petersen, of the same department, has given me vital help in identifying unequivocally the nature of the corrections discussed, for which I am most grateful. I am also greatly indebted to Professor Sir Rudolf Peierls for a very useful discussion on the history of the Peierls effect and for letting me see a preliminary version of his latest writings on the subject.

I am very grateful to Mr Tony Harrison for permission to use the quotation from *Palladas* in Chapter 2. I should also like to record my gratitude to the Galleria degli Uffizi and to S. F. Flaccovio Editori for their generous permission to reproduce copyright material.

I should like now to thank the many people who helped me with comments on various versions of the manuscript. Mr Tim Akrill, Head of Science at Clifton College, had the kindness to arrange for two of his sixth-formers, Finn Spicer and Keith Willmott, to read the whole book. Stretched as they must have been in parts by an almost insuperable task, these young people managed to produce some extremely useful remarks which helped me to improve the readability of the text and I am most grateful for their help. Professor Arnout Ceulemans, who translated the first chapter into Dutch for publication in Leuven, made many useful suggestions and his warm response to the text greatly helped and encouraged me in preparing this volume. Professor Roy McWeeny from Pisa and Dr Peter Herzig from Vienna read the whole of the manuscript and produced a number of detailed comments which were very useful for me in revising the manuscript. Professor Richard Arthur, from Middlebury College in Vermont, made useful comments on the manuscript and I am particularly indebted to him for bringing Anaximander to my attention. Jonathan Maybury, one of my undergraduates at Brasenose, patiently read the whole manuscript and his comments often revealed where I had been too hard on the reader. Miss Alison Roberts, a former under-graduate of mine at Brasenose, very kindly read the manuscript and her expert advice, specially on the solid state section of it, was very useful to me. I am sure that Chapter 1 was greatly improved thanks to my son, Dr Daniel Altmann, whose philosophical eye detected numerous instances of insuffi-ciently precise argument. Finally, I am very grateful to four anonymous readers of the Oxford University Press, who produced a number of perceptive comments and saved me from error more than once. Failures, alas, must surely remain, but if some of the delight of thinking about these subjects permeates to my reader I shall be well satisfied.

Oxford   S. L. A.
October 1991

Credits

Contents

•

1 Ørsted and the principle of symmetry

In the early 17th century the English poet George Herbert tried to describe for us his vision of mankind in a poem entitled *Man*:

Man is all symmetrie
Full of proportions, one limbe to another,
And all to the world besides.
Each part may call the furthest, brother . . .

It is, indeed, very likely that the whole idea of symmetry came to us from the observation of our own bodies, as we can see in many early examples in art. The 5th century BC Greek-Etruscan vase from Tarquinia shown in Fig. 1 presents as well as one can expect in work of this type a symmetry plane, vertical through the ridge of the nose and such that every element of the face on the right of this plane is replicated on the left and at the same distance from it. This type of symmetry is called *reflection* or *mirror symmetry*, because the reflection plane behaves very much like a mirror on which the two halves of the object are reflected. We can see this very clearly in Fig. 2, where you should notice that the ornaments which I have drawn on two corners of the cube are *not congruent*, which means that, if I imagine them detachable, then there is no motion in space that will allow us to superimpose exactly one onto the other. This is precisely the relation between our right and left hands: they are one the mirror reflection of the other, as you can check for yourselves with a high-level mirror, in which you can verify that the specular reflection of your vertically raised right hand (palm facing mirror) is identical to your left hand held upright (palm facing you) underneath the mirror. (This, of course, is well known to everybody, but it is such a strange feature of nature that even those who are utterly convinced always benefit by checking that it is actually true!)

Geometrical symmetry as I have described is extremely clear, but one

Fig. 1. Greek-Etruscan mug in the shape of a girl's head by the potter Charinos, *c.* 500 BC, Tarquinia. From M. Hirmer and P. E. Arias, *A History of Greek Vase Painting*, Thames and Hudson, London, 1962, plate XXXII.

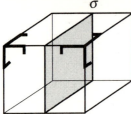

Fig. 2. Reflection symmetry. The grey plane σ is a symmetry plane.

wonders what has it got to do with the properties of real objects. The first difficulty which I shall discuss is that we often mistake icons, that is, the graphical signs which we use to denote objects, for the objects themselves. We see in Fig. 3 a drawing of a man by Leonardo and, on its right, an icon of a man such as, I am afraid, often appears on the doors of gentlemen's lavatories. I do not wish here to get involved with the distinction between a picture and an icon, since the borderline between the two is diffuse: often, indeed, what was meant to be an icon is read as a picture, which should not be surprising because, as anyone with the slightest experience of communication knows, one property of signs is that they often fail to signify. Anyhow, it is clear in Fig. 3 that both the picture and the icon present a symmetry plane. The Leonardo drawing very vividly recalls Herbert's poem: every part on the left may call its opposite on the right brother. Yet, the symmetry displayed is not exact, since the shading, for instance, is not quite identically reflected from the right to the left of the figure. Even more, if we look very carefully at the picture then slight differences appear, and

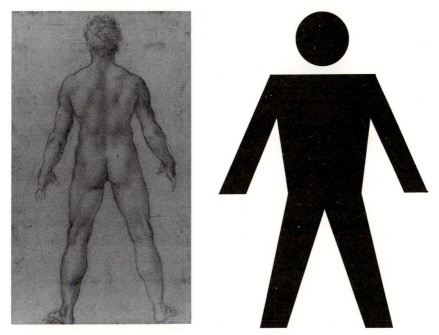

Fig. 3. Objects and their icons. The drawing of a man on the left is by Leonardo da Vinci, from the Royal Library at Windsor.

any serious examination would suggest that despite the superficial appearance of symmetry there might be an underlying asymmetry in the man's body. No such doubts could possibly be raised about the icon, and anyone in need of an appendectomy would be well advised to change his surgeon if the latter is seen studying the icon in preparation for the operation. Such a possibility is, of course, so farcical that even to contemplate it appears to be beyond reason. It is, however, a lamentable but historical fact that more than once physical scientists of the highest repute have gone wrong by proceeding from the symmetry of the icons which they use to that of the objects with which they work, a point which will be illustrated in this first chapter. The second and even less believable possibility of error, which is that of using entirely the wrong icon, will be discussed in Chapter 2.

Before we go any further, it will be useful to discuss a little more the relation between a physical object, its physical model, and its icon.

∎ Physical objects, models, and icons

The reader must not expect a physical object, as so described in this book, to be something that could be grabbed with a pair of pincers. An object for me will be anything that I must use as a scientist in order to describe and analyse the world. Within this meaning, I shall take as my example of a physical object the simplest one possible, namely a force such as the one exerted by a cue on a billiard ball (see Fig. 4*a*). A physical force such as this is always replaced by a *model* of it.

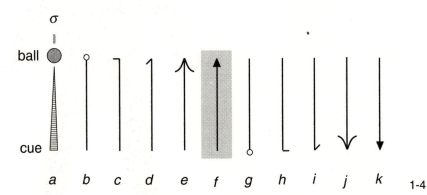

Fig. 4.   Possible icons used to signify a directed segment or a force. The short line labelled σ indicates a plane through it perpendicular to the drawing.

The first thing we do in order to construct our model of a force is to replace the region on which the force acts, which has of course a finite breadth, by an infinitely narrow straight line, called the *line of application* of the force. This line is of course along the cue. The second thing we do in constructing this model is to recognize that rather than a straight line we had better take a segment, the length of which will be proportional to the *magnitude* of the force. The third thing that we must pay attention to is to recognize that not any ordinary segment, even if its length is right, will do: it is one thing to have a given force acting from the cue onto the billiard ball and another to have a force, along the same line of application and of the same magnitude, but now acting from the ball towards the cue. Thus our completed model must be a *directed segment,* that is, a segment the two ends of which are considered to be not equivalent: one end indicates where the force is coming from and the other end where it is going to. In this way our model contains in a rather distilled form the three elements we notice in the physical object: line of application, magnitude, and direction.

Our *model force,* that is, our directed segment, is an entirely geometrical and not a physical concept: segments, like the straight lines of which they are a part, do not exist in nature. So we now have our *physical object* and we have our *physical model*. When we work in physics, we often go one step further and introduce an *icon* to signify or represent the physical model in a graphical way. (The visualization of geometry and its significance in physics is discussed by Reichenbach (1957: p. 97).) So we need an icon to depict a directed segment and it must be clearly understood that there is a lot of freedom in choosing such a depiction. This should be quite clear because an icon is a sign and signs are essentially cultural, that is, they depend profoundly on the particular culture to which one belongs. Thus a gentlemen's lavatory is most often denoted in the UK by means of the icon shown in Fig. 3, whereas in Italy and Central Europe one can still find toilets indicated (albeit rarely) by the symbol 00 (double zero).

We show in Fig. 4*b–k*, various icons, all of which denote exactly the same physical model, namely the directed segment that models the force from the cue to the billiard ball as shown in Fig. 4*a*. Of course, in choosing such an icon any embellishment that establishes a difference between the two ends of a segment will do, and we have displayed in the figure a variety of such embellishments. The icon always used to denote the force in question is of course the one shown in a grey field (Fig. 4*f*) and we would all revolt at the idea of using Fig. 4*k* instead. A culture, I am afraid it must be extra-terrestrial, which had never been exposed to spears or arrows, bless them, would have no trouble at all, however, in happily working with the icon shown in Fig. 4*k*!

I have said that any embellishment that distinguishes between the two ends of a segment would do in order to construct an icon, but I must now qualify this statement. Suppose that we have good reasons to believe that σ in Fig. 4*a* is a symmetry plane: then the icon shown in Fig. 4*c* would not do, because this icon does not show the symmetry. This shows that the icon that we all use, shown in Fig. 4*f*, is to be preferred because it agrees with our known symmetry (this would still be true of the extraterrestrial icon of Fig. 4*k*!).

Once we realize this we can begin to worry for, as we all know, the successful icon of Fig. 4*f*, which most of us will recognize as a *vector* (although strictly speaking it is only the icon of a vector), is the one and only icon that we use for *all* forces and for many other things besides, and the question immediately springs to mind: what shall we do if some of the vectors that we use, unlike the force in Fig. 4, do not have a plane of symmetry through them? I can tell you straightaway what we shall do: we shall have an awful lot of trouble, just as much as the hypothetical surgeon studying his anatomy on the man's icon in Fig. 3.

We are now absolutely ready to begin our pursuit of trouble. Some of you, of course, might sensibly prefer to be spared the agony and the suspense and would like to get the correct answers straightaway. I will not do that, and it is not because I like to spin a long and tricky story, but because there is no solider wisdom than that which is acquired in struggling against trouble. So, we can begin to tell our tale.

▮ Archimedes and the balance

My first physical example will be trivial, as people say, and yet it has caused hot discussion for nearly two thousand years. On the left in Fig. 5 we can see a picture of a balance in which we have two equal weights so that the balance is in equilibrium. I have made this picture rather nice and old fashioned so as to give the idea that, as Leonardo must have had a real man as a sitter for his drawing, I have here a real balance in mind, that is, I am depicting a real experiment in a real laboratory. On the right, we have the icons that the physicists use to describe the experiment: a horizontal bar denoting the arm of the balance, a triangle to denote the fulcrum, and, most importantly, two arrows to denote the gravitational forces.

We can immediately see what is the problem here, because once we draw the icons we appear to have moved from the real world of the laboratory to the abstract world of geometry and, whereas by and large most of us believe that what happens in the laboratory is contingent, geometry has been a notorious repository of self-evident truths, that is, of postulates which were alleged to be self-evident without recourse to experience. Such a view of geometry is of course old hat now, but dogma, even wrong dogma, never dies. We must not forget that Euclid was the first to present a rigorous and apparently consistent mathematical system and that the appeal of Euclid has been a very powerful influence in the way in which scientists and scholars have expressed themselves over the centuries. When Archimedes in the 3rd century BC developed the science of mechanics, he thus borrowed the postulatory–deductive method of Euclid. In Book I of his treatise *On the equilibrium of planes* (see Heath 1897: p. 189), he writes:

> *I postulate the following   1. Equal weights at equal distances are in equilibrium, . . .*

So, if our Minister of Education and Science looks at Fig. 5 he will say: no ladies and gentlemen, I shall not give you a penny to build a balance, just

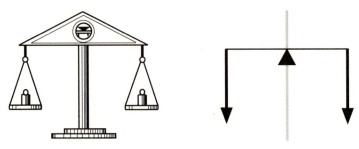

Fig. 5.   A balance and its representation by icons.

draw your icons and with no expenditure at all you will know that the balance must be in equilibrium. No experiment is needed: the truth is self-evident, as recognized by Archimedes. Is it self-evident, is the equilibrium of the balance something we know *a priori*, that is, without recourse to experiment? That this was so appeared to be accepted over several hundred years and the intellectual prop that was created in order to sustain this illusion was a principle of conservation of symmetry which gradually evolved. Before you draw the wrong conclusion, I better say that there was nothing seriously wrong with this principle, although we shall discuss in Chapter 3 how symmetry can actually be broken and thus not conserved. What was wrong was not the principle but rather the way in which it was used. Our first task, therefore, is to try to understand how this principle evolved in practice over the centuries and how it purported to explain Archimedes' 'postulate'.

Early forms of the principle of conservation of symmetry

The first symmetry argument ever used was given by Anaximander of Miletus (*c.* 610–550 BC), whom we know as the first Greek philosopher to have written a book on natural philosophy. Unfortunately, hardly more than a paragraph of this is extant, so one has to follow the references to him given in the later literature. Aristotle in *De Caelo* (see Kirk *et al.* 1983: p. 134) writes:

> *There are some who say, like Anaximander among the ancients, that it [the earth] stays still because of its equilibrium. For it behoves that which is established at the centre, and is equally related to the extremes, not to be borne one whit more either up or down or to the sides . . . so that it stays fixed by necessity.*

We have here the germ of a principle of conservation of symmetry, since the reason why the earth cannot move is that it must maintain the pre-existing equality of its relation to its extremes. This concept was pictorially revived in the Middle Ages by the parable of Buridan's ass. Jean Buridan was rector of the Sorbonne in Paris in 1328 and again in 1340, and he is said to have argued that if you place an ass *exactly* in the middle between two *identical* heaps of hay, he, the ass not the rector, would die of hunger. This, of course, is self-evident, because the ass, assumed to be perfectly rational, would not have any reason whatsoever to move right rather than left, as we can see in Fig. 6. In thus being led into inaction the ass is conserving the mirror symmetry with respect to a plane through the middle of his body which is displayed in the figure.

Fig. 6.   Buridan's ass.

Although the idea of Buridan's ass is important, there is no evidence for the story. It was said that Buridan had written a note in his copy of Aristotle's *De Caelo,* and it is not even sure that the animal mentioned was an ass but, perhaps, less picturesquely, a dog! It was the great Leibniz who put Buridan's idea in a very strong form, appearing at the same time to give his intellectual accolade to Archimedes and his balance. In his second letter to Dr Clarke (1715–1716), Leibniz (see Alexander 1956: pp. 15–16) writes:

> *But in order to proceed from mathematics to natural philosophy, another principle is requisite, as I have observed in my* Theodicy: *I mean, the principle of sufficient reason, viz. that nothing happens without a reason why it should be so, rather than otherwise. And therefore Archimedes being to proceed from mathematics to natural philosophy, in his book* De Æquilibrio, *was obliged to make use of a particular case of the great principle of a sufficient reason. He takes it for granted that if there be a balance, in which everything is alike on both sides, and if equal weights are hung on the two ends of that balance, the whole will be at rest. 'Tis because no reason can be given, why one side should weigh down, rather than the other.*

Leibniz's principle of sufficient reason, as already adumbrated by Anaximander, is one of the major normative tools that we have to this day, but there is something nevertheless in the above quotation that jars our modern eye. This is the expectation that, armed with this principle, we can now happily 'proceed from mathematics to natural philosophy' without the bother, the dirt, and the expense of experiment. In defence of Leibniz, however, it should be said that he understood the dangers of misinterpreting the symmetry of the icon for that of the object. When he discusses Buridan's ass in his *Theodicy* (see Leibniz 1951: p. 150), he points out that Buridan's *Gedanken* experiment is not quite as symmetrical as it looks, because the ass's viscera are not symmetrical and neither are the stars around the ass.

■ The pre-Curie principle of symmetry

There is no question that since the 18th century men and women of science in Europe implicitly accepted a principle of conservation of symmetry, essentially on the basis that if a system had a given symmetry then no effects could be observed that violated that symmetry, very much along Leibniz's line of thought whereby any breaking of symmetry must require a specific cause. (This is why Buridan's ass cannot move!) Strangely enough, it was not until almost the end of the 19th century that a principle of symmetry was formally enunciated, and this was done by Pierre Curie (1894: p. 401). Curie stated his principle in two separate sentences, of which the first is:

Lorsque certaines causes produisent certains effets, les éléments de symmetrie des causes doivent se retrouver dans les effets produits.

When certain causes produce certain effects, the symmetry elements of the causes must be found in the effects produced.

I am afraid I am not entirely fair to Curie in separating this statement from his second one, which we shall consider later, since the two parts of his enunciation are essential in order to appreciate the strength of Curie's ideas. It is not irrelevant, however, that Curie gave pride of place to this particular part of his principle, because he surely was still sufficiently near the intellectual outlook of the early 19th century to have used here language that could perfectly well have come out of the pen of many scientists working at that time. It is for this reason, that I shall call this the pre-Curie principle of symmetry, so as to have a statement of the principle of symmetry as might have been used by scientists like Ørsted in the early 1800s.

I must stress that the complete Curie principle of symmetry, as we shall later discuss, is a perfectly good working tool, still useful and indeed important in our day (see, for example, the excellent treatment of V. A. Kopstik (1983), which summarizes the vigorous work on the principle of symmetry done by the Russian school, also described in Shubnikov and Kopstik (1974)). Indiscriminate use of what I have called the pre-Curie principle, however, had produced a great deal of nonsense for at least a century before it was formally enunciated, since, through Leibniz, it was part of the intellectual baggage of educated Europeans. What is even worse, the pernicious misuse of this putative principle did not cease after Curie published his full enunciation, but it continued until well into the 1950s.

The trouble with the pre-Curie principle is that it appears to work beautifully for the balance, as we illustrate in Fig. 7. It is amply sufficient to

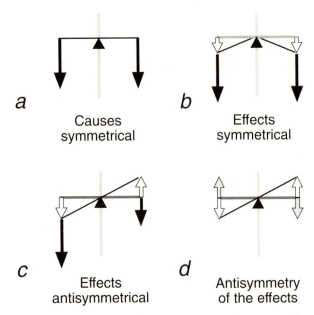

a   Causes symmetrical

b   Effects symmetrical

c   Effects antisymmetrical

d   Antisymmetry of the effects

Fig. 7. Symmetries of the causes and of the effects in the balance. The effects (displacements of the ends of the balance arms) are depicted with outline arrows. Forces are depicted with black arrows and reflected effects are the same outline arrows but filled in grey. Figure *d* is merely used to help to explain why, in *c*, the effects are stated to be antisymmetrical. The conventions used here are valid for all the figures in this chapter: causes are given in black and effects in outline, whereas all reflected objects (whether causes or effects) are in grey.

use icons, which show that the causes are *symmetrical* with respect to a symmetry plane (Fig. 7*a*). Thus, the effects, which are the displacements of the two ends of the balance's arm, given with outline arrows in the figure, must be symmetrical as shown in Fig. 7*b* and not *antisymmetrical* as depicted in Fig. 7*c*. Let me explain what we mean by this statement. In Fig. 7*d* we show the effects, that is the displacements of the ends of the balance's arm, as outline arrows, whereas the reflection of these effects on the symmetry plane of the balance is given by the same type of arrow but filled in grey. It is thus clear that the reflected effects entail, at each end, the opposite motion to that at that end, and it is this change of sign of the effect which is indicated by the word *antisymmetric*. Because the effects must be symmetrical, and because the (finite) symmetrical effects of Fig. 7*b* can be ruled out, unless the balance breaks, the only possibility compatible with the symmetry that the effects must have is for the two white arrows in Fig. 7*b* to vanish, that is for the balance to be in equilibrium.

We have thus achieved a remarkable piece of Thatcherite efficiency, namely to get something for nothing: we have obtained an experimental result without performing an experiment! All this reeks of metaphysics and it is difficult to believe that the scientific community ever took this type of reasoning seriously. Only as recently as 1958, however, David Park (1958: p. 215), wrote: 'It has occasionally been said that the only *a priori* propositions of physics relate to symmetries, and this may well be true, . . .'. He, of course, strongly qualifies this statement with a warning that new results must come from experiment, but he nevertheless witnesses to what had been an indisputable trend amongst scientists.

That the analysis of the Archimedes principle via the pre-Curie symmetry principle is nothing if not intellectual rubbish must be manifest to us, for it is entirely based on the assumption that the symmetry of icons is identical with the symmetry of the physical objects that they signify, a point which must be seriously investigated, as we have already seen. The dubious metaphysics here, on the other hand, did not escape the notice of the followers of Poincaré's conventionalism (see Giedymin 1982), as we shall now discuss. Looking back at the balance in equilibrium (Fig. 7*a*) we must ask the following question (see Nagel 1961: p. 55): how do we know here that the forces on both sides of this balance are equal, *except* through the fact that the balance is in equilibrium? Thus, we do not have to postulate metaphysically that the balance is in equilibrium because the forces are equal. Rather, we use this analysis to *conventionally* define what we mean by equal forces. Metaphysics has thus been neatly replaced by *conventionalism*. I should like to stress, however, that conventionalism in this form is nothing more than disguised metaphysics, since it is accepted as part of the convention that the null experiment exists in nature. However obvious it may seem that this must be the case, that is, that balances can be in equilibrium, we shall see that this is a *contingent* and not a logically necessary fact, even for empty balances. (All this will be cleared up later on; see p. 34. The concept of force, on the other hand, merits deeper discussion: see Reichenbach (1957: p. 22).)

The cosy world of the symmetry conservationists was shattered in 1820 by Ørsted's discovery of the electromagnetic interaction, which did not mean, of course, that they instantly died in shame. On the contrary, their ideas as enshrined in what I have called the pre-Curie symmetry principle, were still influential for another 140 years at least. So, the time has come to talk a little about that remarkable man Ørsted.

Ørsted and the electromagnetic interaction

Hans Christian Ørsted is one of the most unusual scientists of the 19th century, although he himself would have rejected the then non-existent appellation of scientist: he regarded himself as a literary man. He was born on 14 August 1777 in southern Denmark, the son of the village apothecary, who was of course nothing much more than a humble herbal-medicines dispenser. Poor, Ørsted senior may have been, but he (or Mrs Ørsted) must have had some remarkable genes since their two sons Hans Christian and Anders (later Minister of State) so distinguished themselves that local patrons arranged for them to go to university in Copenhagen. Hans Christian gained there a degree in pharmacy in 1797 and was awarded a doctorate in 1799, the subject of his thesis being Kant, a point which is not without significance in understanding his later work.

The years of the Napoleonic wars are somewhat mixed for Ørsted. He works at the Lion Pharmacy in Copenhagen and he falls under the infuence of the *Naturphilosophie* movement then strong in Central Europe as a trend associated with the spread of romanticism. His association with the German physicist Ritter and with the Hungarian chemist Winterl causes his reputation to suffer: the problem with the *Naturphilosophie* movement was that they were inspired rather than critical and thus prone to support crankish ideas, such as a belief in astrology. Their idealistic philosophy came largely out of Kant and, with him, they gave preponderance to the concept of force over that of matter. Matter is diverse but forces can all be thought to be of the same nature: this led them into a concept of the unity of nature, which, however mystical it may have been in its foundations, had the most profound influence in the later development of science and mathematics during the 19th century. Although *Naturphilosophie* was largely a Central European movement, it did reach England, mainly through the poet Coleridge, who went to Germany in 1798, and of whom we shall hear more in Chapter 2.

The *Naturphilosophie* movement was not alone in propounding a dynamical theory of matter in which forces are the underlying feature of nature. In the second half of the 18th century the Croatian Jesuit R. J. Boscovich had proposed an atomic theory in which atoms were a manifestation of forces. Whereas this theory had a strong following in Britain, from Davy to Faraday, there is a fundamental difference: the *Naturphilosophie* partisans were quite uninhibited about making what at the time appeared to be metaphysical arguments. (Remember that atoms were then inaccessible to observation.) Davy and Faraday, on the other hand, were reluctant even to reveal their belief in Boscovich's atomic ideas and did their best to convey public personae as pragmatic scientists (Williams 1965).

Electricity was the high technology of the Napoleonic era, if something without any application whatsoever can be called a technology. But the production of electrical batteries and the display of their properties was the wonder of the age. Already in 1798 Ritter, guided by the *Naturphilosophie* principle of the unity of nature, had equated electricity with chemical affinity (see Williams 1965). In the same way, it was this principle that moved Ørsted, as early as 1812, in a volume entitled *Ansicht der chemischen Naturgesetze*, published in Berlin (see Oersted 1812), to propose: 'one should try if one could not bring about an action on the magnet as such by electricity in its very latent form' (Ørsted 1920: Vol. 2, p. 148). Ørsted was quite good at building acid batteries and yet it took him eight years to perform the experiment that made him famous, changed the world of physics, and created the conditions for the technological revolution brought about by electricity later on in the century.

Why eight years? The complete answer to this question we shall never know, but it is highly probable (see the Appendix) that during those years Ørsted performed repeatedly the wrong experiment until, in 1820 (the same year when he discovered the alkaloid piperine), the result came: Ørsted and his audience at a lecture observed that a magnetic needle is deflected when electricity is passed through a nearby wire. This, in fact, is the phenomenon we now observe everyday when we see the hands in dials of electrical instruments moving to indicate the passage of current.

The date of the discovery is given by most historians as April 1820 and, as we shall see, it is highly relevant. It seems to me, however, that the evidence for this data is tenuous and that January to March 1820 cannot be ruled out and might, in fact, be more plausible dates (see the Appendix). Ørsted was not normally given to let the grass grow, yet he did not publish his result until 21 July 1820. From his notebooks at the Royal Library in Copenhagen (see Fig. 13, which reproduces a page from 15 July 1820), we know that, during this period, he not only explored very carefully and very deeply the consequences of his discovery but also went on repeating the wrong experiment, the one that he had incorrectly expected would succeed. It is likely that his delay in publishing was due to this, but when he did publish his paper, he wrote it in Latin as well as in Danish and sent it to six major European towns. Never before in the history of European science could such excitement be felt, and the two or three months that followed 21 July 1820 are probably without parallel in the whole century. Literally and metaphorically, they were electrifying.

We can fortunately follow the story week by week in the minutes of the Monday meetings of the Academy of Sciences of Paris (see Institut de France 1916; the page references that follow all refer to this volume). On 4 September Arago communicated the results of the experiments made by 'M. Œrstedt' (so unknown was he that his name was not properly spelt until 13 November!) and repeated by de la Rive at Geneva (p. 83). One week later, 11 September, Arago himself reproduces the Ørsted experiment and reads his memoir to the Academy (p. 90). On 18 September Ampère reads a paper including results additional to those of Ørsted (p. 94). On 25 September Ampère demonstrates experimentally to the Academy the interaction between two electrical currents (p. 95). On 9, 16, and 30

October Ampère communicates results of new experiments, including the magnetization of steel by the electric current and, at the last of these meetings, Biot presents a memoir on the physical laws governing the experiments made as a result of Ørsted's discovery (pp. 97–99). On 6 November Arago describes new results about magnetization by ordinary electricity (p. 100). Finally, on 4 December (p. 108) Ampère reads his famous paper on the mathematical laws for the attraction and repulsion of electrical currents in which he provides the mathematical theory of the electric field. London, meanwhile, is not asleep: already on 10 October Sir Humphry Davy had informed Faraday of the Ørsted effect (Williams 1965: p. 151) and the latter had immediately repeated the experiment. Without all this the fundamental discoveries by Faraday of electrical induction could never have happened. The unknown Danish pharmacist Ørsted became famous.

Fame, however, did not bring undiluted glory: already three months after the discovery, Professor Gilbert, on publishing a German translation of Ørsted's paper in *Annalen der Physik*, of which he was the editor, wrote: 'What all investigations and cares did not produce was given by chance to Professor Oersted of Copenhagen.' Ørsted retorted with copious denials, in various journals in 1821 (*Danish Royal Academy, Journal für Chemie und Physik, Journal de Physique*), in his Autobiography (1828), and in an article in English in the *Edinburgh Encyclopaedia* of 1830. He died on 9 March 1851, so famous that his funeral in Copenhagen was reported to have been attended by the Crown Prince, ambassadors, ministers, and, unbelievable as it might be today, 200 000 people! (Dibner 1961).

Then, on 30 December 1857 lightning struck: Christoffer Hansteen, a Norwegian physicist, wrote a letter to Faraday (quoted in Jones 1870: Vol. 2, p. 389) saying that the discovery was indeed accidental, that Ørsted at his lecture had performed his failed experiment, which of course did not work, but that, at the end of the lecture, he changed the orientation of the electrical wire and it was then that the needle moved.

We shall see that the orientation of the electric wire with respect to the magnetic needle is of enormous importance to our discussion and yet I think that too much had been made, by his contemporaries as well as by Ørsted himself, about whether he did or did not place the wire in the correct orientation deliberately: the mere fact that he kept on and on trying to see whether there was any interaction between a current and a magnet is something for which we must always remember him, because it is almost certain that the great electrical experts of the time were never even near doing that (see the Appendix). Thus, even if Ørsted had lied when claiming that he had deliberately placed the wire in the correct orientation, I do not think it diminishes his achievement by an ounce. I am inclined to believe that his account of the experiment was not entirely candid, and that he was perhaps all the more human for that. Competition, envy, jealousy, prejudice against the man from outside, both geographically and scientifically (he could never feel comfortable in front of the real professionals like Arago) were pressures that few could have endured easily. On the other hand, when he had the phenomenon in his hands, his grasp of the problem was first class.

The most detailed and most quoted account of his discovery as given by Ørsted appears in his article on 'Thermoelectricity' in *The Edinburgh Encyclopaedia*, the holograph manuscript of which, in English, is in the Royal Library at Copenhagen but has never been published, so that I have analysed the relevant paragraph in the Appendix. Analysis of this section of the article in its various versions does nothing to explain why Ørsted so suddenly changed his mind as to the experiment he wanted to perform and, on the contrary, gives the impression of a feeble attempt at explaining what appears most likely to have been accidental. I thus suggest that Hansteen's account is preferable to that of Ørsted. (Stauffer (1953) produces evidence that Hansteen could not have witnessed the discovery, but his argument hinges on accepting April 1820 as the date of the discovery, whereas I show in the Appendix that January and February of that year, when we do know that Hansteen was in Copenhagen, cannot be ruled out.)

▌ The Ørsted paradox

We come at last to an analysis of the Ørsted experiment, and in doing this analysis we shall try to think as Ørsted himself may have done. Do not forget that he had written a thesis on Kant, that he was most probably familiar with the Leibniz principle of sufficient reason, and that the language that Curie later adopted for the part of his symmetry principle which we have quoted was precisely the language that Ørsted would have used. It is thus no anachronism to imagine Ørsted using what we have called the pre-Curie symmetry principle.

The experimental set-up is simplicity itself, as depicted in Fig. 8*a*: all you need is a fairly thick conductor to carry the current **i** and a magnetic needle freely pivoted around its centre. (Ørsted himself, for a time, used thin platinum wires, because he thought that any effect that might be observed would be associated with the emission of light and, perhaps, heat. He later, however, realized that a thick wire was best.) So as to have the needle initially at rest, the vertical plane depicted is chosen always to coincide with the magnetic meridian. Naturally, it is easier to discuss the experiment by using icons (as Ørsted himself largely did; see Fig. 13) and we do this in Fig. 8*b*. In Fig. 9, again using icons alone, we describe two alternative configurations which the experimental set-up can take, with the current

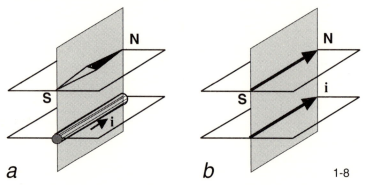

Fig. 8. The Ørsted experiment. The vertical shaded plane is the plane of the magnetic meridian, N and S being the north and south poles respectively.

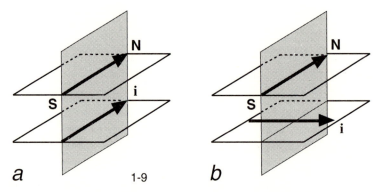

Fig. 9. The parallel (a) and perpendicular (b) configurations for the Ørsted experiment.

parallel (*a*) or perpendicular (*b*) to the meridian plane where the needle rests. (There are, of course, many intermediate alternative configurations, but consideration of these two extreme cases suffices.)

In order to apply the pre-Curie symmetry principle to this problem, we must first look at the symmetry of the causes, which we study in Fig. 10. In order to see what is the symmetry of the magnetic needle with respect to the vertical (meridian) plane, we imagine in Fig. 10*a* the needle moved away from the plane, so that we can easily form its mirror image under reflection, which is the grey arrow. If we now assume that the needle moves towards the plane, both arrows eventually coincide: this means that reflection on that plane leaves the arrow invariant, and because of this we say that the magnetic needle is *symmetrical* with respect to the meridian plane. (Remember please that all the discussion in this section is merely a historical reconstruction of Ørsted's possible mental processes and that *nothing here is necessarily true*! The truth will eventually emerge, of course, but not yet.) Exactly the same argument is valid for the arrow representing

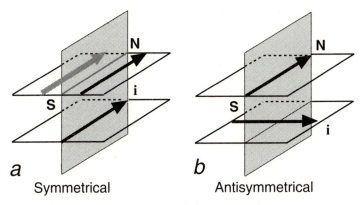

Fig. 10. The symmetry of the causes in the parallel (a) and perpendicular (b) configurations of the Ørsted experiment. The grey arrow is the reflection on the vertical plane of the black arrow.

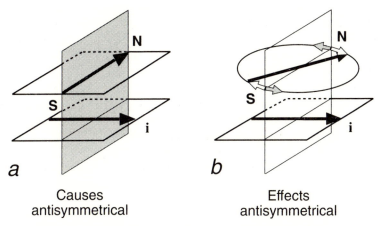

Fig. 11. Naïve application of the principle of symmetry to the perpendicular Ørsted configuration. The effects (displacements of the needle ends) are given with outline arrows, and their reflections are in the same type of arrow but filled in grey.

the current, so that we conclude that in this configuration the causes are symmetrical with respect to the meridian plane. It is easy to see in the same manner that the second experimental configuration, represented in Fig. 10b, is *antisymmetrical*, that is, that the so-called causes experience a change of sign on reflection through the meridian plane. Armed with the principle of symmetry, we can now easily predict what the experimental results, if any, will be, which we do in Figs. 11 and 12.

In the perpendicular configuration the causes are antisymmetrical (Fig. 11a), so that any possible effects must also be antisymmetrical (Fig. 11b). We support this last statement in the same manner as was used for the balance: the effects are given by the outline arrows which indicate the possible rotation of the magnetic needle ends. The outline grey arrows, which are the reflections of the effects through the vertical plane, are such that the displacement of each end of the needle is reversed, thus showing that the effects are antisymmetric. In this configuration, symmetry is conserved after displacement of the needle and a non-null experimental result may be expected. Notice that when we say here that symmetry is conserved we really mean that the symmetry type discovered, which is antisymmetrical, is conserved in going from the cause (initial configuration) to the effect.

Let us now look at the parallel configuration. Here the causes are symmetrical (Fig. 12*a*) and the effects (Fig. 12*b*), just as in Fig. 11*b*, are antisymmetrical. Symmetry is not conserved and a null result must be observed, since it is only the null result that will be symmetrical, as the causes are. Notice that in all this we are following exactly the same principles that successfully explained the Archimedes balance. Nature, alas, is not so kind to us in this case. (It would be more accurate to say that there is a limit to the amount of human nonsense that Nature can put up with!) The perpendicular configuration is the one that Ørsted unsuccessfully tried, probably for some years (see the Appendix), and it is the one that he went on trying even after the first successful experiment of 1820, as Fig. 13 shows, reproduced from his notes for 15 July 1820. The successful experiment, against all apparent common sense, was realized in the parallel configuration of Fig. 12, in which, in fact, Ørsted and his audience actually observed precisely the same type of finite displacement of the magnetic needle as illustrated in Fig. 12*b*. It is difficult for us now to imagine the consternation that this result must have produced in those who were

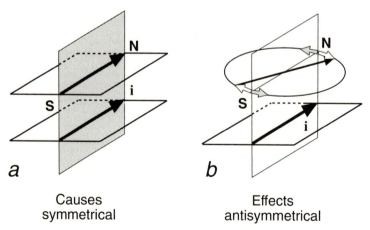

a	*b*
Causes	Effects
symmetrical	antisymmetrical

Fig. 12.   Naïve application of the principle of symmetry to the parallel Ørsted configuration. The effects (displacements of the needle ends) are given with outline arrows, and their reflections are in the same type of arrow but filled in grey.

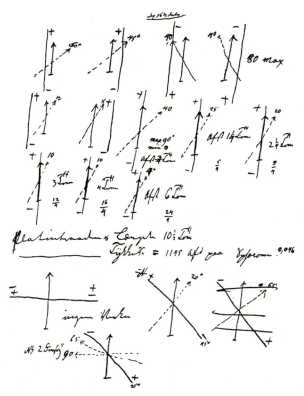

Fig. 13. Page of Ørsted's laboratory notes for 15 July 1820 (from Ørsted 1920, Vol. 1, p. LXXXII). Notice: (i) the use of icons to depict the magnetic needle; (ii) a sketch, near the bottom-left, of the experiment in the unsuccessful perpendicular configuration.

thoughtful enough to realize its paradoxical significance. Fortunately, however, we have the recorded reaction of one of the clearest minds in physics from the 19th century. Ernst Mach (see Mach 1893: p. 27) wrote in 1883:

> *Even instinctive knowledge of so great logical force as the principle of symmetry employed by Archimedes, may lead us astray. Many of my readers will recall, perhaps, the intellectual shock they experienced when they heard for the first time that a magnetic needle lying in the magnetic meridian is deflected in a definite direction away from the meridian by a wire conducting a current being carried along in a parallel direction above it.*

Notice, incidentally, how Mach refers to the principle of symmetry, although its explicit formulation by Curie was still eleven years away. It is not too difficult to put one's finger on the reasons for the Ørsted paradox, since the whole of the argument on which the false conclusions are based depends on identifying the symmetry of the icons with the symmetry of the objects which they denote. In the case of Buridan's ass nothing short of a dissection would convince us that the celebrated ass is not symmetrical (as Leibniz reminded us). It would be foolish of course to try to examine the internal structure of the pieces of metal of which the wire and the needle are made. We do notice, however, that they are both depicted by vectors and it is therefore worthwhile looking carefully into the nature and the symmetries of such objects.

∎ The nature of vectors

Not even the word *vector* existed in Ørsted's time, since it was not invented until 1846 by Hamilton, but Ørsted and his contemporaries already understood some of their features. In order to produce a convincing analysis of Ørsted's experiment, I shall try to use ideas that he or any of his contemporaries could have used. We can thus agree that for us, as already suggested, a vector will be a physical directed segment, by which we mean that we must have a quantity that is associated with a rod, say, and that the two ends of the rod must be *physically* distinct. If we can assume that the rod is now infinitely thin, we can first replace the object in question (that is, the physical rod) by its model, a directed segment, and then represent the latter by its icon, the usual arrow, the head and the tail being distinct in order to denote the difference between the two ends of the rod. The rod thus becomes a directed segment in this limiting case, and this is what we mean by a vector, at least for the time being.

If we want to produce such constructs, it is just as well to deal with mechanical problems, since they were well understood even at the beginning of the 19th century. Our first example of such a construct is shown in Fig. 14. In part *a* of this figure we show a reference frame, which we take to be stationary, with a scale which is fixed in this reference frame. We also see in this figure a rod, parallel to the *x* axis. This rod lies on a huge turntable centred at *C* (not shown and normally stationary) which can rotate around this point without affecting the scale shown along the *x* axis. If the rod is perfectly uniform and it is not moving, there is no way in which we can tell one of its ends from the other.

Let me explain this. Of course, we could construct a directed segment from x_1 to x_2, but this geometrical directed segment will do nothing to make the two ends of the rod distinct: it will not be sufficient to say 'the end at x_1' to identify unequivocally one end of the rod because if we flip over the turntable by 180° the rod is in exactly the same physical state as before

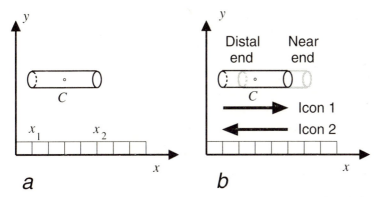

Fig. 14. A model for polar vectors. The position of the displaced rod during the motion is indicated in grey.

(that is, we do not even notice that the flip has taken place!) and the end we had in mind will have changed name.

In Fig. 14*b*, on the other hand, we are in business. What we have done is to impress on the rod a velocity parallel to the positive branch of the *x* axis. The rod thus moves on the turntable (that is why it must be huge, so that the rod does not get out of it) and therefore it moves relative to the fixed frame. We can now define uniquely the two ends of the rod by means of convenient names. One end we shall call the *near end* and it is such that, during the motion, its coordinate *x* never coincides with a value that at some previous time has been assumed by any other point of the rod. The other end, which does not satisfy this property, will be called the *distal end*. Relative to the defined fixed frame, the distinction between these two ends is *intrinsic* in the sense that as long as the motion is not perturbed we can always identify the near end, say. We can see that this is so by flipping the turntable by 180° (we must assume that this can be done without upsetting the motion of the rod). The near end will now appear to the left of the figure but it will still be the near end: it boldly moves where no other point of the rod had been before! That is, the name 'near end' is as fixed to the end in question as if it had been painted in red, whereas before, in Fig. 14*a*, the name 'end at x_1' did not label one specific end of the rod. (Topographical names cannot distinguish identical particles!)

Having now convinced ourselves that a rod possessed of a velocity behaves like a directed segment, we can represent it by a vector icon, as in Fig. 14*b* (icon 1), but it must be carefully understood that the choice of this icon is purely a matter of convention: as we have already said (see Fig. 4) a culture which had never been exposed to bows and arrows or spears would have found the icon 2 in Fig. 14*b* just as good, since all that we want is to make a distinction between the two ends of the rod.

We have now obtained our first vector. We shall see in a minute that there is another entirely different type of vector, for which reason we have to qualify the name of the one just invented. We shall call it with the now current name of *polar vector*, although we shall also sometimes refer to it as a *velocity-like vector*.

There is another way in which we can make the two ends of a rod distinct and this is simply by rotating the rod about its own axis, as we illustrate in Fig. 15. As far as the man on the right is concerned, the rod is rotating *counterclockwise*: his *left* hand is lowered following the motion. The woman on the left, instead, has to lower her *right* hand to follow the motion: she sees the motion *clockwise*, as indicated in her balloon. Because the two ends of the rod are now, from this point of view, distinct, we are entitled to denote such an object by a vector icon. The disposition of the head of the arrow is just as conventional as it was for the polar vectors. The usual rule is to place it at the end of the rod from which the rotation is seen as counterclockwise, as we do in the picture. Such vectors are now called *axial vectors* and we shall also refer to them from time to time as *rotation-like vectors*.

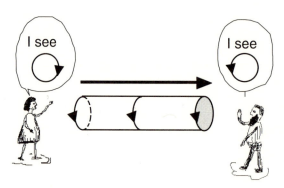

Fig. 15.   Rotating a rod makes the two ends of a rod distinct: in the limiting case of an infinitely thin rod this leads to a directed segment, that is, a segment the two ends of which are distinct.

This example of an axial vector illustrates very clearly the distinction between a physical object (rod rotating), its model (rod rotating with the rod diameter infinitely narrow), and its icon (arrow).

We must immediately notice, however, a bit of a problem: in dealing with polar and axial vectors we are using precisely the same icon for two entirely different objects, namely both polar and axial vectors. This is an unfortunate practice, but this is what people have always done and it would be foolish at this late stage in history to try to change this practice (although some hopeful people have tried). It is pretty obvious on the other hand that we must be very careful not to get mixed up and we can readily imagine that the origin of the Ørsted paradox might easily lie here. To see whether this is the case we must first study the symmetry properties of vectors.

❚ The symmetry of vectors

It will be sufficient for us to find out the symmetry properties of vectors, whether polar or axial, with respect to reflection planes, and we shall consider two cases for such planes, namely when they are respectively parallel or perpendicular to the corresponding vectors. In reading the figures which we shall use, we must distinguish clearly between the object and the image fields, that is, between the object and its reflection through the plane used: in all our figures everything in the object field will be black and everything in the reflected field will be grey. We consider polar vectors first, in Fig. 16. If the rod is parallel to the reflection plane (Fig. 16*a*), it is clear that when it moves up its image in the reflected field also moves up, which means that the corresponding icons point upwards in both cases (Fig. 16*b*). If, instead, the rod is perpendicular to the reflection plane (Fig. 17*a*) and it moves up away from this plane, then its reflected image moves down, also away from the plane, as shown in the figure, and correspondingly for the icons (Fig. 17*b*). We notice that so far the

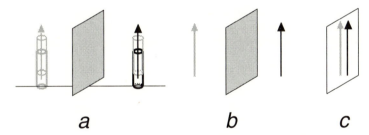

a   *b*   *c*

Fig. 16. Polar vectors are symmetrical with respect to reflection through a parallel plane. The object field is always on the right of the reflection plane shown and all reflected objects are given in grey.

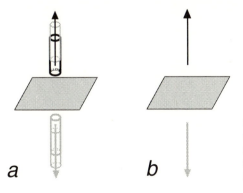

a   *b*

Fig. 17. Polar vectors are antisymmetrical with respect to reflection through a perpendicular plane. The object field is always above the reflection plane shown and all reflected objects are given in grey.

behaviour of the icons is precisely what would be expected of decent Indian arrows. We say, from Fig. 16*b,c*, that polar vectors are symmetrical with respect to reflection through a parallel plane (notice, in fact, that if the object vector lies on the plane, as would be the limiting case of Fig. 16*c*, its image coincides with itself and the vector is thus invariant). Figure 17*b* shows that a polar vector is antisymmetrical (changes sign) with respect to a reflection through a plane perpendicular to the vector.

All these results are so trivial that one would be tempted to obtain them by reasoning from the icons alone (which is indeed what many people have done instinctively for centuries), but this would be hopelessly wrong, as we shall see when dealing with axial vectors. In order to discover transformation properties, it is absolutely essential to deal, as we have done here, with the real objects that the icons denote.

We treat axial vectors in Figs. 18 and 19. In Fig. 18*a* we have a rod rotating counterclockwise (as seen from the top) in the object field, whereas its reflected image rotates clockwise (also as seen from the top). In both cases the vector icons must be placed with their heads at the end of the rod from which, from outside the rod, the rotation is seen as counterclockwise, and when we do this we get the surprising result that the icons are flipped over under reflection through a parallel plane, a behaviour which is even more shocking when the icons alone are displayed, as we do in Fig. 18*b*. Naturally, if we let the object vector in this figure move until it reaches the reflection plane, we can see that it changes sign, that is, that it is antisymmetrical with respect to any plane that contains the vector (Fig. 18*c*).

Let us now consider the reflection of an axial vector through a plane perpendicular to it (Fig. 19). Here the top shaded base transforms into the corresponding (also shaded) base at the bottom, without change of the sense of rotation (always as seen from above the top of the object rod). It is clear that the corresponding vectors do not change sign under this reflection. We have thus proved that axial vectors are *antisymmetrical* with respect to a plane parallel to them, and *symmetrical* with respect to a perpendicular plane (this latter result follows from Fig. 19*b*). Although this behaviour is counterintuitive from the icons, you must remember that what really matters is what the corresponding objects do.

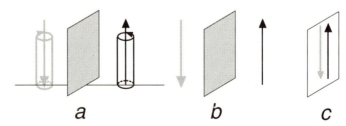

a   b   c

Fig. 18. Axial vectors are antisymmetrical with respect to reflection through a parallel plane. The object field is always on the right of the reflection plane shown and all reflected objects are given in grey.

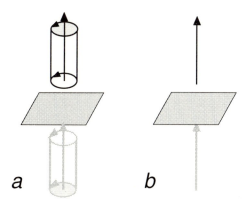

a   *b*

Fig. 19. Axial vectors are symmetrical with respect to reflection through a perpendicular plane. The object field is always above the reflection plane shown and all reflected objects are given in grey.

We should expect the symmetry behaviour of an object to be an intimate part of its nature and thus that axial vectors must be entirely different from polar vectors, although they can successfully masquerade for them via their ambiguous icon. In modern terminology, in fact, polar vectors are vectors *tout court* and axial vectors are not vectors at all but rather *antisymmetrical tensors of rank two*, of which more in our second lecture. (To say that a vector is a directed segment is thus a gross simplification, since both polar and axial vectors satisfy this criterion.)

It should now be clear that it was injudicious of Ørsted to assume that he knew the symmetry of his 'causes' as he, anticipating Curie, would have said. It is also pretty clear that what we have called the pre-Curie symmetry principle is a very dangerous tool, because it invites one to assume that the symmetry of the causes is known in some unspecified way or, even worse, that it can be considered self-evident. This is nothing except metaphysical equivocation, since it should be clear from our discussion of vectors that the symmetry of things is not always that which meets the eye. In order to analyse properly the Ørsted experiment, we need a better symmetry principle.

Aquinas, Curie, and the principle of symmetry

As I shall show in a moment, Curie had a perfectly clear idea that symmetry had to be derived from experiment and not to be dreamt up. This being the case, it is very unfortunate that his enunciation of the principle of symmetry started with the statement on p. 9, since this statement is useless unless the symmetry of the causes is known, and there is nothing in that statement as such to give us an indication of how this symmetry might be determined. It thus appears to be sheer metaphysics.

Before we state Curie's perfectly correct answer to this problem, it is useful to have a look at a much earlier attempt to get rid of metaphysics when relating causes to effects. Naturally, this was done by a theologian, St Thomas Aquinas, in the 13th century, a good six hundred years before Curie. In the *Summa Theologica* (see Aquinas 1856: Vol. 1, p. 31, Book 1, Q4, Art. 2) St Thomas had in fact put the whole matter on what appears to be a sound empirical basis:

> *Primo quidem per hoc quod quidquid perfectionis est in effectu, oportet inveniri in causa effectiva. . . . Manifestum est enim quod effectus praexistit virtute in causa agente.*

> *First, any perfection which occurs in an effect must occur in its efficient cause. . . . For it is manifestly true that an effect virtually pre-exists in its active cause.*

As Feynman (1963: Vol. 1, p. 52-12) says: 'We have, in our minds, a tendency to accept symmetry as some kind of perfection.' It is thus admissible to substitute 'symmetry' for 'perfection' in Aquinas' enunciation and we immediately have a normative principle for *determining symmetry experimentally*, which is what a good symmetry principle should do for us. Notice that what Aquinas says is this: observe effects experimentally, determine their symmetry, and then, from this observation, you can deduce the symmetry of the physical objects involved in producing the observed effects (the 'causes'). One could in this way, in principle, design experiments to determine the symmetry properties of physical objects.

I shall call the statement that 'any symmetry observed in an effect must also occur in its cause' the *Aquinas principle of symmetry*, but it must be abundantly stressed that, if used at all, this principle must be applied under strictly controlled conditions, of which more later. This is so because, admirable as the intentions of the Aquinas principle are, it unfortunately fails as a general symmetry principle for the simple reason that, in certain circumstances, an effect can be far more symmetrical than the cause that produces it: physical systems have the uncanny property of symmetrizing very rapidly asymmetrical perturbations which they may experience. (Most physical scientists are familiar, for instance, with the principle of equipartition of energy, which spreads out the energy received in one particular degree of freedom of a system into all its degrees of freedom.)

Because the symmetrization of perturbations is so important in understanding the principle of symmetry, I illustrate in Fig. 20 what must be its simplest possible example. We have in *a* a homogeneous circular steel disc which is being heated by a flame. This is the cause of the thermal expansion of the disc and, as shown in *a* this cause has one and only one plane of

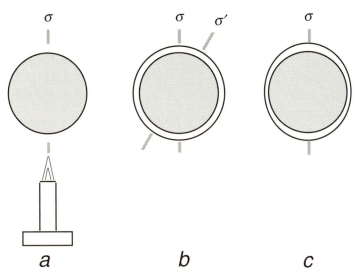

Fig. 20.   Expansion of a circular disc. The cause or perturbation (heat source) has only one symmetry plane (*a*), whereas the effect, that is, the expansion of the disc (*b*) has the full circular symmetry with an infinite number of reflection planes, such as σ′, in addition to the original plane σ. In *c* an effect is shown which has the same symmetry, that is, one and only one symmetry plane, as the cause in *a*. The planes σ and σ′ are perpendicular to the drawing through the lines shown.

symmetry, σ. In *b* and *c* we show two possible forms of the effect, that is, of the expansion which the disc experiences if it is heated some time and then separated from the flame. In *c* this expansion has only one plane of symmetry, just the same as the original cause. Clearly, however, owing to the high thermal conductivity of the steel, the temperature in the disc will become uniform in an extremely short time after it is separated from the flame, which will mean that in practice the effect observed will be the isotropic expansion of the disc shown in *b*. This effect, of course, has a lot more symmetry than the original cause, since any plane through the centre of the disc, like the plane σ′ shown, is a symmetry plane, as is characteristic of circular symmetry. Observation of the circular symmetry of the expanded disc in *b*, used in conjunction with the Aquinas principle, would lead to the wrong conclusion that the cause must also be circularly symmetrical, which is clearly in contradiction with *a*, where we do not have an infinite number of symmetry planes but only one. Notice that, on the other hand, the pre-Curie principle of p. 9 works, since the symmetry plane of the cause (the vertical plane just mentioned) still subsists in the effect.

It is rather sad to have to say that wise St Thomas has not given us a completely general rule: from the existence of symmetry in an effect it is not possible to deduce that the same symmetry exists in the corresponding cause. It is only the *asymmetry* of an effect that licenses us to assert that the same asymmetry must exist in its cause. For instance, if c were observed, in which the original circular symmetry of the steel disc is reduced destroying all symmetry planes except one, we could then correctly infer that the cause must be similarly asymmetrical with respect to the circle. If, on the other hand, b is observed, as is the case in practice, it would be wrong to conclude from the circular symmetry of the effect that the cause must also be circularly symmetrical.

It might be worthwhile warning the reader about a possible fallacy here, since it is tempting to believe that the higher the symmetry of a system is, the higher is its order. Thus, our statement that the symmetry of the effect might be higher than that of its antecedent state, might appear to violate the second principle of thermodynamics. High symmetry, however, is associated with high and not with low disorder. Consider, for instance, the disordered melt from which a crystal grows, both the melt and the crystal being assumed of infinite extent. Whereas the melt admits of every possible rotation and every possible reflection as a symmetry operation, the crystal possesses only a subset of these symmetry elements. Thus, the disordered melt is *more symmetrical* than the ordered crystal! In the same way, the expanded disc with full circular symmetry is more disordered than the distorted disc with only one symmetry plane.

We can now safely go back to our symmetry principle. The requirement that the asymmetry of the effect must entail the same asymmetry in the cause is nothing other than an application of the principle of sufficient reason. Why does this principle not allow us to infer, à la Aquinas, symmetry in the cause from symmetry in the effect? My way of looking at it is as follows. We must accept in science that the relation between cause and effect is mediated by dispositions. (The idea of disposition is, I am afraid, somewhat delicate philosophically: see Goodman (1965: p. 40).) Macroscopic effects, such as the thermal expansion of a crystal, depend on microscopic dispositions which very often symmetrize the perturbations received (causes). Thus, symmetries of observed effects may not coincide with the symmetries of the perturbations that originated them.

It is this fact that was clearly recognized by Pierre Curie when he enunciated the second part of his principle of symmetry. This follows immediately the sentence which we transcribed on p. 9 (see Curie 1894: p. 401):

> *Lorsque certains effets révèlent une certaine dissymétrie, cette dissymétrie doit se retrouver dans les causes qui lui ont donné naisance.*

> *When certain effects show a given asymmetry, this asymmetry must be found in the causes which have given origin to them.*

It should be stressed that, in contrast to what I have called the pre-Curie principle of symmetry, this enunciation does not depend on any *a priori* assumptions about the symmetry of the causes and that it is thus an honestly experimental normative physical tool. As I have already suggested, however, it is regrettable that Curie did not give this important statement first place in the enunciation of his principle of symmetry, since it is this part of the principle which has a primary experimental significance. Indeed, the first part of Curie's principle (that is, our pre-Curie principle) can only be used after sufficient experimentation has established beyond any possible doubt, through the second part, the symmetries of all the causes efficient in a given experiment. It must also be said that, in adopting that order, Curie has invited the partial quotation of the first part of his principle, as a 'principle of symmetry' as I, like many others before, have done on p. 9, with pretty disastrous results, since it invites the use of assumed and in practice often untested symmetries of the physical objects which are the antecedents or causes in a given physical experiment.

Having given this warning about the misuse of the first part of Curie's principle of symmetry, that is, of what hitherto we called the pre-Curie principle, it is important to recognize that, properly used as explained above, it is a valid and useful principle. Because it requires that the symmetry of the causes be maintained in the effects, this statement is a principle of *conservation of symmetry* which is to this day one of the great principles of physical science and which can be regarded as a corollary of Leibniz's principle of sufficient reason. If the first part of Curie's principle can be considered a principle of symmetry conservation, its second part is instead a principle of conservation of asymmetry, but it must be remembered of course that these two principles are applied in entirely different ways.

It should be clear from our discussion of the symmetrization of effects that deducing the symmetry of the cause from that of the effect, à la Aquinas, is risky. If one is dealing with simple systems, that is, systems with a sufficiently small number of degrees of freedom, where the symmetrization of asymmetric perturbations cannot be expected, then the Aquinas principle can nevertheless be tried as a quick expedient, but it is always prudent to check it properly by application of the full Curie principle.

▌Ørsted: no more paradoxes

Let us see how this works by imagining we are planning, now properly, an experiment on the lines imagined by Ørsted. We sketch in Fig. 21 two possible configurations for the two vectors **μ** (magnetic needle) and **i** (current). It is reasonably clear (and, in any case, it will be proved later on) that, whatever the nature of the vectors **i** and **μ**, the only possible symmetry

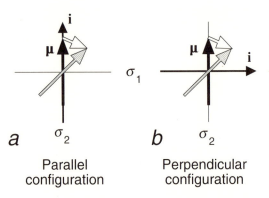

Fig. 21. The parallel and perpendicular configurations of the Ørsted experiment and the nomenclature of the reflection planes. The hatched arrows indicate the position of the magnetic needle as a presumptive result of the experiment, whereas the outline arrows indicate the displacement of the head of the magnetic needle, that is, the presumptive observed effect.

a   Parallel configuration   *b*   Perpendicular configuration

planes that the antecedents of the experiment might have are the planes labelled σ_1 and σ_2 in the figure. We must first of all study the symmetry of the effect (rotation of the needle) that we could observe. It is easy to see that, in either of the two configurations, any rotation of the needle breaks any possible symmetry with respect to the planes σ_1 and σ_2. This is so because the vectors that denote the effects, which are the outline arrows in the figure, are not invariant with respect to the two planes in question. Notice also that the plane through the displaced needle is not a plane of symmetry since $\mathbf{i}$ and $\boldsymbol{\mu}$ are not invariant with respect to it. We thus conclude, from the Curie principle as just stated, that:

> Any rotation of the needle possibly observed after switching the current is incompatible with the existence of a symmetry plane in the system.

Armed with this result our programme of work is as follows. We do not know at this stage what type of vectors $\boldsymbol{\mu}$ and $\mathbf{i}$ are, but we can easily consider all possible cases for them. In each case, from the vector type assumed for $\boldsymbol{\mu}$ and $\mathbf{i}$, it is not difficult, from the rules given for axial and polar vectors, to find out whether a symmetry plane exists or not in the parallel or perpendicular configurations. If such a plane exists then the result in the box above indicates that the magnetic needle cannot move out, whereas if no symmetry plane remains in the system for the given choice of vector types then an effect may be observed, that is, the needle may deflect out of the meridian plane. When this is done we shall find that neither of the two configurations can be ruled out from the point of view of a possible effect, despite the conclusion obtained from the naïve application of the pre-Curie principle that the only possible result would be obtained in the perpendicular configuration. Moreover, comparison of this analysis with the experimental result will give us useful information about the possible types of vectors that we have in the experiment and thus about their

Table 1. Symmetries in the Ørsted experiment.

			Parallel				Perpendicular			
			σ_1	σ_2	Effect	Exp.	σ_1	σ_2	Effect	Exp.
1	μ	polar	A	S	0		A	S	+	
	i	polar	A	S			S	A		
2	μ	axial	S	A	0		S	A	+	
	i	axial	S	A			A	S		
3	μ	polar	A	S	+	OK	A	S	0	OK
	i	axial	S	A			A	S		
4	μ	axial	S	A	+	OK	S	A	0	OK
	i	polar	A	S			S	A		
1	2	3	4	5	6	7	8	9	10	11

A = antisymmetric; S = symmetric; 0 = no effect; + = possible effect; OK = effects agree with experiment.

symmetries: the Ørsted effect will thus be seen as a symmetry-determining experiment.

All this work is done by means of Table 1, which some readers might find somewhat intimidating, but I shall go through its construction step by step.

We must first of all consider all possible combinations of the vector types for μ and i, which we do by treating the four cases numbered 1 to 4 in the table. In cases 1 and 2 both vectors are assumed polar and axial, respectively, and in cases 3 and 4 one of the vectors is assumed polar and the other axial.

Once these working assumptions are made, the symmetry of the vectors with respect to the planes σ_1 and σ_2 is easily determined from the rules which we have given in Figs. 16 to 19, and we enter it in the columns headed by the symbols which denote the corresponding planes. Consider case 3 in the parallel configuration as an example. The polar vector μ is antisymmetric (A) with respect to its perpendicular plane σ_1 and it is symmetric (S) with respect to the plane σ_2 that goes through it. The vector i, being axial, is symmetric (S) with respect to its perpendicular plane σ_1 and antisymmetric (A) with respect to its parallel plane σ_2.

We have now accounted for the first five columns of the table, plus of course columns 8 and 9 as well (see the numbering of the columns at the bottom of the table). We can now use the rule in the box above in order to predict in each case whether, given the assumptions made about the vectors, an effect might be observed. Take, for example, case 1. The entries for this case in column 5 being both S, show that σ_2 is a symmetry plane. We can see at once that for this choice of vector types the parallel configuration, by keeping a symmetry plane, is not compatible with a possible effect, as it follows from the rule in the box. In order to indicate this we enter '0' under 'Effect' in column 6 of the table. In the perpendicular configuration neither σ_1 nor σ_2 is simultaneously a symmetry plane for both vectors, so that our rule permits rotation of the needle and we enter '+' under 'Effect' in column 10.

Except for columns 7 and 11 our table is now complete. What we have done so far is the work Ørsted should have done in planning his experiment. If he had done this he would have anticipated that an experimental effect could be observed just as well in the parallel as in the perpendicular configuration, so that he need not have wasted perhaps eight years, neither need Mach have been astonished by the result. In order to corroborate what I have just said, all that we need is to look for '+' signs in columns 6 and 10. From column 6 we see that an effect is possible in the parallel configuration in cases 3 and 4, that is, when the vectors μ and i are both of different types, one axial and the other polar. From column 10, conversely, an effect is possible in the perpendicular configuration in cases 1 and 2 when the vectors μ and i are both polar or both axial.

Let us now compare with experiment. We know, thanks to Ørsted's work, that an effect is observed in the parallel configuration and that no effect is observed in the perpendicular one. Our predictions are entirely the opposite of this result in cases 1 and 2 but agree precisely with the experimental results in cases 3 and 4, as indicated by the entries in columns 7 and 11.

This analysis shows that the Ørsted experiment is an experimental observation of the symmetry behaviour of the vectors μ and i. Of course, there is an element of ambiguity in this experimental result, because we cannot discriminate between cases 3 and 4 and thus we cannot say which of the two vectors is the axial and which is the polar vector. Since, however, a current must be associated with a linear motion, that is, a velocity, it is not unreasonable to give it a velocity-like, that is, a polar, vector so that μ must therefore be an axial one.

Having determined experimentally that one of the two vectors involved in the experiment must be axial and the other polar, we can go further, because we can see what must happen if the relative positions of these vectors are exchanged. In Fig. 22a we show the result of Ørsted experiment as so far described, with the magnetic needle above the current. In Fig. 22b we have reflected the two vectors on the horizontal plane, assuming for convenience that it is μ which is the axial vector, as we have suggested. Because this vector has changed is sign, so also must the sign of the effect, the vector δ, change, since otherwise symmetry would not be conserved. This is one of the sixty or so experiments which Ørsted performed in July 1820 and from which he concluded that the 'electrical conflict' associated with the current must entail a circular motion (the circles shown in the

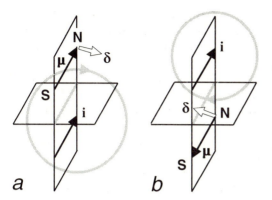

Fig. 22. The parallel configuration of the Ørsted experiment (*a*) and the change in the effect (outline arrow **δ**) obtained when the magnetic and the electric-intensity vectors are reflected through the horizontal plane of the figure (*b*). The vertical plane is the magnetic meridian and N is the (of course fixed) direction of the north magnetic pole of the earth.

figure are such that the displacements of the north pole of the needle are always tangential to them). As he wrote at the end of his paper:

we may conclude that a circular motion likewise occurs in these effects.

Of course, Ørsted visualized the 'electrical conflict' as some sort of perturbation caused by the current in the space around it, which we replace today with the concept of the magnetic field of the electrical current.

One last remark about the Ørsted experiment. We have seen that it is sensible, from the experiment, to take the current and magnetic vectors as polar and axial respectively. This implies that magnetism must be related to some rotational phenomenon, as axial vectors are. It is interesting that Ørsted himself intuited this, but dared not make these thoughts public. In a scratched page of his manuscript on 'Thermoelectricity' in the Royal Library at Copenhagen (p. 84) he writes:

The electromagnetism reveals to us a world of secret motions, no doubt worthy of mathematical inquiry ... It is not improbable that even magnetism involves certain rotations ...

Of course, we know in modern work what those rotations are, since magnetism is simply recognized as a manifestation of the electron spins, all in parallel coupling in the case of a ferromagnetic material. So we see that the great Ørsted story ends by leading us to the electron spin, precisely as the Hamilton story in Chapter 2 will end, although the trail that we shall follow could not be more different.

■ Archimedes again

Let us now turn our attention again to the Archimedes balance. Since this is a simple mechanical system we can tentatively try the Aquinas principle, which I would like to do since I want to illustrate fairly quickly that the state of equilibrium of the balance, which is a symmetrical effect and thus not a useful input for the Curie principle, is not one that can be assumed to exist without experimental facts.

Consider the balance in equilibrium for equal forces (Fig. 23a). The null effects observed are symmetrical with respect to σ and thus, from the Aquinas principle, the causes must also be symmetrical. This means that the forces must be symmetrical with respect to σ, that is, that they are polar vectors. (Remember that it is only polar vectors that are symmetrical with respect to a parallel plane.) If we were in a region of space where forces are axial vectors (Fig. 23b), the balance would not be in equilibrium ever (except of course for null forces). This is so for the following reason. As it is clear from Fig. 23b, the causes in this hypothetical case are antisymmetrical and thus the first part of the Curie principle (which can now safely be applied) determines that all effects must be antisymmetrical, so that a state of equilibrium can never be observed, unless the forces are null. (In this latter case the causes are symmetrical and thus a null, symmetrical, effect is possible.)

The important experimental fact about the balance is that its state of equilibrium, that is, the null effect, *exists for non-null forces*. Once we ascertain this result we use it for defining equal forces, which is just as exciting as choosing the metre as the unit of length: the conventionalist interpretation of the balance thus misses the one fundamental experimental fact while emphasizing what is only a trivial choice.

The application of Aquinas' principle for the balance can easily be checked by Curie, but in a more roundabout way: we must start with a single force on the right of the balance, say, and thus with an anti-symmetrical effect. We deduce that the cause, that is, the moment of that force, must be antisymmetrical, whence the force must be a polar vector. (We have to assume here, and in all this work on the balance, that the vector denoting the distance from the point of application of the force to the fulcrum is a polar vector: this is as far as we must go in assigning *con-ventionally* the nature of polar vectors to position vectors in space.) Once it

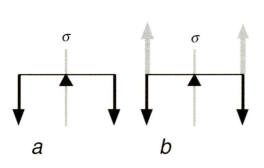

Fig. 23.   The balance and the symmetry principle. In a the forces are polar vectors. In b the forces are assumed to be axial vectors, which are antisymmetrical with respect to the reflection plane σ, as shown by their reflections through that plane, given with grey arrows. Since the null effect observed in the equilibrium configuration of the balance is symmetrical, such an effect is incompatible with the assumed axial vectors as possible causes.

is thus established *experimentally* that forces are polar vectors, then the first part of Curie's principle will determine the equilibrium of the balance for equal forces.

Notice from the above discussion that, just like in the Ørsted experiment, there is an element of choice in the Archimedes problem since we have to conventionally define position vectors to be polar. From the point of view of physics this is a choice that could not worry anybody. It must be recognized, however, that there is often an element of convention in assigning symmetries from experiments. What is very important on the other hand is to recognize that null experiments are the major experimental facts and as such contingent: their existence or otherwise is not subject to convention!

130 years after Ørsted: parity non-conservation

One would have thought that the lesson learned from the Ørsted paradox would have stopped people from ever again making the same mistake. This, alas, was not so, as we shall see on discussing the following experiment (Wu *et al.* 1957), which is now so well known that it is part of history. We describe this experiment in Fig. 24. We have a crystal of ^{60}Co at very low temperature in the centre of a solenoid, represented in the picture by a loop. The magnetic field of the solenoid orients the spins of the nuclei in a direction normal to the plane shown in the figure (this is why low temperature is needed, to avoid this alignment of the spins being disturbed by thermal motions). Because the spins behave like axial vectors (being rotation-like quantities) they are symmetrical with respect to the plane

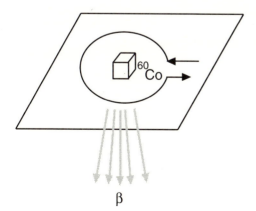

β

Fig. 24. The experiment of Wu *et al.* The solenoid is represented by a single winding.

shown in the figure (compare with Fig. 19). Therefore, the whole system is symmetrical with respect to this plane and no effects asymmetrical with respect to it could be observed. (Of course, we are reasoning in the pre-Curie principle style!) ^{60}Co decays and produces β rays (electrons), and the faulty reasoning described would lead us to expect that the electron count would be identical above and below the plane shown in the drawing. This experiment could have been done many years before, but it was not tried because the result seemed obvious: when it was done, however, the electron count observed was different on either side of the plane. This result was stated in terms of the property of *parity*, rather than that of symmetry with respect to a reflection plane as we have done. The system in Fig. 24, after taking account of the nuclear spin properties as discussed, *appears* to be symmetrical with respect to inversion at the origin, a property which is called parity. The result of this experiment was thus that parity may not be conserved, a find that was to be one of the most important ideas of the 1960s.

The question with the experiment of Wu *et al.*, or rather with the long delay in actually trying to do it, is of course that, again, preconceived rather than experimental ideas about symmetry were used in such a way that an unwarranted preconceived result was expected.

Because this is the fundamental conclusion of this lecture, let me summarize it again. The idea that symmetry is a geometrical property of objects and that it can therefore be determined by geometrical and not by experimental means results often from confusing the icon of an object, which is indeed a geometrical entity, with the object itself. Symmetry must be determined experimentally and it is a very sad fact that the crucial experiments required for this purpose, such as the Ørsted effect and the balance, are never discussed as symmetry-determining experiments. It cannot be overemphasized that just as one is prepared to determine the resistance of a wire experimentally, so, given a new physical concept or model, its symmetry properties must be determined by the careful application of the Curie principle in well planned experiments.

Although Leibniz, through the principle of sufficient reason, must be considered as the father of the symmetry-conservation principle, his assertion that this is what is needed in order 'to proceed from mathematics to natural philosophy' (which he repeats *twice* in the quotation given on p. 8) is if not wrong at least seriously incomplete. Of course, there is a two-way trade between the two disciplines, but at the crossroads of science it is the opposite direction that must be followed: we must go from natural philosophy, that is, from experiment, to mathematics. It is only in this way that the icons provided by mathematics can be clothed with their correct transformation (or symmetry-behaviour) properties.

Let me therefore summarize the modern form in which Curie's principle of symmetry must be used. The asymmetry of observed effects must be assigned to their corresponding causes or antecedents. In this way the symmetry properties of the antecedents can be determined. If and only if the symmetry properties of all the antecedents intervening in a given experiment are thus known beyond any reasonable doubt, then their symmetries can be used in the principle of symmetry conservation (Curie's first principle) to determine the symmetries of the corresponding effects.

In concluding this chapter I should like to pay homage to Hermann Weyl, who, in his marvellous book on symmetry, said in half a page (almost) everything which I have said here (Weyl 1952: p. 20).

∎ Appendix: Ørsted's discovery

The most quoted account of the discovery is that given by Ørsted in his article on 'Thermoelectricity' in *The Edinburgh Encyclopaedia* (Brewster 1830, quoted from Ørsted 1920: Vol. 2, pp. 357–8; the braces, the meaning of which will later be explained, are mine):

In composing the lecture in which he was to treat of the analogy between magnetism and electricity, he conjectured, that if it were possible to produce any magnetic effect by electricity, this could not be in the direction of the current, since this had been so often tried in vain, but that it must be produced by a {lateral action. This was strictly connected with his other ideas; for he did not consider the transmission of electricity through a conductor as an uniform stream, but as a succession of interruptions and re-establishments of equilibrium, in such a manner, that the electrical powers in the current were not in a quiet equilibrium, but in a state of continual conflict. As the luminous and heating effect of the electrical current goes out in all directions from a conductor, which transmits a great quantity of electricity; so he thought it possible that the magnetical effect could likewise eradiate. The observations above recorded, of magnetical effect produced by lightning, in steel needles not immediately struck, confirmed him in his opinion. He was} nevertheless far from expecting a great magnetical effect of the galvanical pile; and still he supposed that a power, sufficient to make the conducting wire glowing, might be required. The plan of the first experiment was, to make the current of a little galvanic trough apparatus, commonly used in his lectures, pass through a very thin platina wire, which was placed over a compass covered with glass. The preparations for the experiment were made, but some accident having hindered him from trying it before the lecture, he intended to defer it to another opportunity; yet during the lecture, the probability of its success appeared stronger, so that he made the first experiment in the presence of the audience. The magnetical needle, though included in a box, was disturbed; but as the effect was feeble, and must, before its law was discovered, seem very irregular, the experiment made no strong impression on the audience. It may appear strange, that the discoverer made no further experiments upon the subject during three months; he himself finds it difficult enough to conceive it; but the extreme feebleness and seeming confusion of the phenomena in the first experiment, the remembrance of the numerous errors committed upon this subject by earlier philosophers, and particularly by his friend Ritter [itals], *the claim such a matter has to be treated with earnest attention, may have determined him to delay his researches to a more convenient time. In the month of July 1820, he again resumed the experiment, . . .*

There are three points which require discussion in this account. The first is the question of the date. Stauffer (1953) bases his dating of the discovery on this account: July 1820 minus the three unexplained fallow months give April 1820 as the date of the discovery, which is quite relevant, since he gives convincing evidence that Hansteen could not have been at Copenhagen in March and April 1820. Thus, in accordance to Stauffer, Hansteen's evidence can be dismissed as mere hearsay.

Although, with Stauffer, April 1820 is most often given as the date of the discovery, I find this date open to strong suspicion. There are, in fact, three *earlier* accounts of the discovery by Ørsted which do not agree with this dating. They are as follows in chronological order: (1) Right at the beginning of his July 1820 paper Ørsted says that the experiments were done 'last winter'. (2) In his 1821 paper (Oersted 1821, quoted in Ørsted 1920: Vol. 2, p. 224), he changes the date to the spring of 1820. (3) In Ørsted's autobiography from 1828 (see Stauffer 1957: p. 50) he says:

The idea first occurred to him in the beginning of 1820 while he was preparing to treat the subject in a series of lectures on electricity, galvanism and magnetism. He had set up his apparatus . . . Burdened by daily routine for several months . . . In July he renewed the experiments . . .'

It would be natural to take the very first account as to be the most likely to be correct, as done by Whittaker (1951: Vol. 1, p. 81), which would suggest January, February, and possibly March as likely dates for the discovery. This, if coupled with the third account, would favour January and February, whereas March again is just about compatible with the second account. In the light of these datings, I think that April is improbable, January or February are likely, but that March cannot be ruled out. Thus, Hansteen's evidence, admitting that he could not have been in Copenhagen in March or April, cannot be ruled out of hand as hearsay. Having said this, I confess that there is one point that worries me about

Hansteen's evidence (Jones 1870: p. 390), for he says that the effect observed at the discovery was a great one 'almost at right angles with the magnetic meridian'. Not only is this in contradiction with all Ørsted's statements about the discovery, but it is also difficult to explain, since it is almost certain that Ørsted was using thin platinum wires at that time.

The second point which arises from the 'Thermoelectricity' quotation given above is Ørsted's reference to his thought 'that if it were possible to produce any magnetic effect by electricity, this could not be in the direction of the current, since this had been so often tried in vain'. This is most important: it is clear in this paragraph of 'Thermoelectricity' that Ørsted is trying to provide an explanation of why he thought that a lateral action could be in operation in the process and he creates the impression in the lines just quoted that the negative evidence obtained, by other people, persuaded him to try a different line of thought. (Obviously, if he himself had been performing the perpendicular experiment always, it would require some explanation why he had changed his mind!) I am afraid that I take the view that Ørsted is not entirely candid here, since it is almost certain that the only one person who could have been trying the experiment at all was Ørsted himself. We have this on the evidence of the great Ampère, who knew everything that had been done in this field. Writing to Roux-Bordier, 21 February 1821 (quoted by Williams 1965: p. 142) he asserts: 'You are quite right to say that it is inconceivable that for twenty years no one tried the action of the voltaic pile on a magnet'. And he then explains that this was nevertheless the case because of everybody's blind acceptance of Coulomb's hypothesis about the nature of magnetic action. This accords very well with the facts as we know them and it explains why it took a man like Ørsted, who was at the margin of the major ideas on electricity at the time, to find the effect.

The third point is the crucial one, namely how far is it plausible that Ørsted could have foreseen the existence of a lateral action. It must be said straightaway that if Ørsted did foresee this, such a thought alone would have been one of the most striking ideas for a century or more. It must be clear from our discussion of the symmetry of the experiment that all the training that Ørsted had must have moved him towards the perpendicular configuration, as we have already shown his notebooks testify. It must also be understood that the Ørsted effect is the first instance in the history of humanity where a transverse force (in modern terminology, a tensorial interaction) was found: until then all known forces acted along the line of action of the vectors that originated them. It appears to me that Stauffer (1957), who defends Ørsted (as if he needed defence: his achievement was undeniable), accepts a little too readily the possibility of such an intellectual feat.

Let us now look for the internal evidence in 'Thermoelectricity' on this question of the lateral action. In assessing how far Ørsted found it easy to explain his claim of having predicted it, we are fortunate in having the holograph manuscript of 'Thermoelectricity' in the Royal Library at Copenhagen (*Ørsteds Saml.* 23, 2°). There are in it (page 16 of the manuscript) some small emendations and departures from the text quoted above. What is rather interesting is that the section of the published text shown in braces in the quotation given on p. 37 was pasted over in the manuscript. When, at my request, the conservation department of the Royal Library detached the correcting strip, a previous version was found on its back, so that we can now follow Ørsted's thoughts through three versions of the same sentence in the manuscript, and this is the crucial sentence in which he claims that the idea of a lateral action came naturally to him. The first version chronologically, written on the manuscript folio itself, said:

lateral [effect] action, likewise as the calorific and the [lo] luminous effects of the violent electrical current in a resisting medium, [acts] consists principally . . .

and it then goes on along this line. The second version, on the back of the pasted strip, says:

lateral action. As he did not consider the electrical current [as an uniform] [illegible] *[stream like water] but as a [continuous] succession of ruptures and reestablishment of equilibrium*

The third version, on the front of the pasted strip, which is the one remaining uncorrected in the manuscript, is:

lateral action. This was not inconsistent with his other ideas; than he [consider] did not consider the [electrical current as] transmission of . . .

Notice that it takes Ørsted three trials (in a manuscript which is otherwise fairly free from corrections) to try to justify that the lateral action 'was not inconsistent with his other ideas' and that, even then, he changed this in the printed version (see the main quotation on p. 37) to a fourth and stronger statement:

this was strictly connected with his other ideas;

I am not claiming that this is in any way conclusive evidence, but it shows how hard it was for Ørsted to justify his allegedly inspired guess, thus making it more difficult to accept the intellectual feat claimed by him. In any case, I take the view that the discovery of the effect in whatever way, whether by design or by accident, was in itself a remarkable achievement since the mainstream of European thought at that time was going off in an entirely different direction.

2 Hamilton and rotations

I was promised a horse but what I got instead
was a tail, with a horse hung from it almost dead.

Tony Harrison, *Palladas: Poems*, 47, 1975

On looking at Fig. 1 some of my readers will ask themselves: what on earth has the legend of St George and the dragon got to do with symmetry? The answer to this question is, I am afraid, nothing at all. Yet, I show this picture here not just for the fun of it but because it will teach us a salutary lesson about icons, and icons, as we have seen, have to be well understood if one does not want to talk nonsense about symmetry. The way most people nowadays will read this picture is like this: the artist has depicted in it three icons, those of a maiden, a knight, and a dragon, in order to represent the legend in which St George slew a dragon and thus saved a maiden in need. It is not impossible that even the painter of this picture, Raphael, thought that he was doing that. Insofar, however, as he was following a Byzantine iconographic tradition, he was actually depicting a region (Cappadocia, of which the maiden is the icon, maidens often being used as icons for towns) subjugated by its heathen population (the dragon), which was converted by a saint (the knight). It was, in fact, this type of picture that *created* the legend of St George and the dragon, rather than *depicting* this legend as we now read it. The point I want to illustrate with this example is that it is quite possible not just to mistake an icon for the object it is supposed to represent, as we have seen in our first chapter, but also to mistake it for an entirely different object, which is going to be the story treated in this chapter.

The symmetry operations I shall be concerned with are rotations, and a beautiful example of rotational symmetry is given by the mosaic from the Villa Filosofiana in Sicily shown in Fig. 2. If we disregard the inner roundel of the mosaic, a rotation of the whole picture by 45° ($2\pi/8$) will cover the figure, that is, will not change its aspect in any way. Notice also that the pattern of the mosaic is such that no other symmetry operations remain: the shading of the decoration is such that it destroys all symmetry planes. Naturally, we have here not just one rotation by $2\pi/8$ as a symmetry

Fig. 1.   St George and the Dragon. Drawing by Raphael at the Cabinet of Drawings and Prints, Galleria degli Uffizi, Florence, no. 592 E. This is a cartoon for the oil painting at the National Gallery of Washington.

operation, but also those by $4\pi/8$, $6\pi/8$, $8\pi/8$, and so on until $16\pi/8$, which is a rotation by 2π. This is a very special type of operation, which, like the number zero, appears to do nothing and in fact does nothing, but it is very crucial in order to get your mathematics right. Just as the use of the number zero was the crucial step that allowed the Arab arithmeticians (who got it from the Hindus) to revolutionize mathematics, so the invention of the operation that does nothing, or the *identity operation*, changed our views on symmetry. Hence, the symmetry of our mosaic is called *cyclic*,

Fig. 2.   A floor mosaic from Villa Filosofiana, Piazza Armerina, Sicily. A. Carandini, A. Ricci, and M. de Vos, *Filosofiana, The villa of Piazza Armerina. The image of a Roman aristocrat at the time of Constantine.* S. F. Flaccovio, Palermo, 1982. Plate XVIII, p. 43.

because all the eight operations that we observe in it are given as a cycle of the same operation, the rotation by $2\pi/8$, repeated 1, 2, ..., 8 times. We shall hear more about this symmetry in our third chapter.

The type of rotation we have illustrated in Fig. 2 is not the most general rotation we can imagine, because the axis of rotation is perpendicular to the object being rotated rather than having an arbitrary orientation. To depict the most general rotation that we can imagine, all that we need is an axis of rotation **n** and a vector **r** at an angle to it, as illustrated in Fig. 3. Of course, we have arranged that the point of application of the vector **r** to be transformed, that is, the tail of this vector, is on the same line as the axis **n**. Quite clearly, rotation by α (counterclockwise) around **n** transforms the vector **r** into the vector **r**′ on the surface of a cone with axis **n** and of which **r** is one of its generators. All this is as clear as pure water is. Life, however, is not as simple as that, and at the end of this chapter we shall learn why this is so.

The first person who dealt in a very carefully mathematical way with rotations was Hamilton and he clearly had Fig. 3 in his mind but he did not call α the angle of rotation but rather *twice* the angle of rotation. He could not even bear to call Fig. 3 a rotation *tout court*, but he had to embellish the name into that of a *conical transformation*. You may think that Hamilton was mad, but before you jump to conclusions you must appreciate that Hamilton was inventing not only a theory of rotations but also the very concept of a vector and, as you will see later, what he thought he was representing with a vector icon was something entirely different, just as happens to us when we look at Raphael's picture and we read a dragon-icon rather than a heathen-icon.

Some of you readers may feel by now that we have enough trouble in our lives than to go and spend time thinking about other people's mistakes: and thus you will feel like closing this book. If this is so, please listen to my entirely disinterested advice (my royalties are paid to me even if you burn this book) and go on reading, because Hamilton's mistake was the result of an excess and not of a lack of wisdom, and some of the latter will, I hope, stick to us as well if we make a little effort. We may recall, besides, that the Archimedes example of the balance in Chapter 1 appeared to be ridiculously simple and yet it took us to the core of some of the major

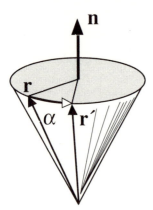

Fig. 3.   A rotation by the angle α about the axis **n**, to be denoted with the symbol $R(\alpha \mathbf{n})$.

discoveries in physics in the 19th and 20th centuries. We shall likewise see that the analysis of rotations upon which we shall now embark will lead us to remarkable results about the nature of space which have paved the way this century for the understanding of a very important property of the electron, namely that of possessing spin.

So, let us start our story, first with a few words about the remarkable Hamilton.

▌Sir William Rowan Hamilton

Young William was not, of course, born a knight, neither was he born wealthy, for his father was a not-very-successful attorney in Dublin, where Hamilton was born on 3 August 1805.* Because the family could not easily support the children, William was sent at the age of three to his paternal uncle James, a vicar at Trim. It was soon clear that the child was a prodigy. At seven it was said by competent authorities that his command of Greek was on par with that of top university graduates, and his mental arithmetic could have made a living for him on the entertainment circuits. At 17 he produced a paper accepted by the Royal Irish Academy on caustics, which was to revolutionize the design of telescopes and which formed the basis for much of his later work. At the age of 22, before he graduated at Trinity College, Dublin, Hamilton was made Astronomer Royal of Ireland, and he was knighted at 30.

As with Ørsted, a grasp of Hamilton's work depends crucially on an understanding of his philosophical ideas and it is no coincidence that he shared with Ørsted the same world outlook, in a way which would have been unexpected of a scientist or mathematician in, say, England or France. Not only was he a Kantian like Ørsted, but he also embraced the ideas of the *Naturphilosophie* movement. In a *Chapter on Astronomy* (Graves 1882: Vol. I, p. 501) he uses, already in 1831, when he was 26, language that could have come straight from Ørsted's pen: 'The design of physical science is . . . to learn the language and interpret the oracles of the universe.'

Like Ørsted, Hamilton was a literary man and moved in literary circles. He was the godfather of Wordsworth's grandson and he was asked by Lady Wilde, but he declined, to be Oscar Wilde's godfather also (Ellmann 1987: p. 16). In 1832 Hamilton travelled specially to Highgate in England in order to meet Coleridge, with whom a lifetime friendship and correspondence followed. Remember that Coleridge was the prophet of the *Naturphilosophie* in England and like many of the followers of this philosophy he

*For more details of the historical references to Hamilton and to Rodrigues given here, see Altmann (1986, 1989). Good modern biographies of Hamilton are Hankins (1980) and O'Donnell (1983).

was also a Pythagorean, having like Pythagoras a mystical belief that numbers held the key to the universal mysteries, a view which of course had an extraordinary appeal for Hamilton and which would become central in his later work. They regarded the world, as it were, as a great onion that had to be peeled off layer by layer by pure thought until the inner core would reveal the truth of the universe. And Hamilton considered it necessary, even more, considered it his moral duty, to keep his vigilant eye on these multiple layers all at once in order to discover this hidden truth. So, whereas Fig. 3 appears to be of trivial clarity for us, this was not so for him, who had to discharge moral obligations unknown to lesser mortals. To understand these moral obligations, we must start the story right at the beginning, and please do not be impatient if the first page or two of this saga do not appear to have anything at all to do with rotations. When we discuss imaginary units immediately afterwards, we shall already begin to make contact with rotations and everything will fall into place eventually.

■ Complex numbers

It was Hamilton who first showed that the complex numbers form an algebra of couples. Let me explain what we mean by this. A complex number A is defined as follows

$$A = a1 + Ai, \qquad a \text{ and } A \text{ real}, \qquad (1)$$

where the numbers 1 and i (imaginary unit) multiply in accordance with Table 1.

Table 1. Multiplication table for the complex number units.

	1	i
1	1	i
i	i	-1

The way in which two complex numbers

$$A = a1 + Ai, \qquad B = b1 + Bi, \qquad (2)$$

multiply is very well known: you multiply the right-hand sides as if they were ordinary bionomials and then you substitute the results of Table 1 for the products that entail two of the complex units 1 and i:

$$AB = ab - AB + (aB + bA)i. \qquad (3)$$

This means that complex numbers can be defined as an *ordered pair* of real numbers or a *couple*, denoted as

$$A = [a,A], \qquad B = [b,B], \tag{4}$$

and that then their product, from (3), is also a couple:

$$AB = [ab - AB, aB + bA]. \tag{5}$$

Notice in fact that the right-hand side of (3) contains a real number and an imaginary term, the latter being the product of the real number $aB + bA$ times the imaginary unit. The rule that we have used in order to go from (2) to (4) will also yield (5) from (3). The fact that the product of two complex numbers (couples) also gives a couple, validates the use of the definition of complex numbers as couples.

Since complex numbers are a generalization of real numbers, Hamilton realized that real numbers must be written as a special case of the complex couples and he chose the obvious form

$$a = [a,0], \tag{6}$$

which clearly satisfies the condition that the product of two reals must be a real number. Keep an eye on this simple idea because, although absolutely correct here, it will lead us to one of the villains in the story that follows.

∎ The imaginary unit

What we are going to do now appears to be as innocent as an unborn baby and, yet, it is a clue that will lead us to the second of our villains. Remember that Hamilton had to relate whatever he did to the world at large and he wanted therefore to understand the meaning of the multiplication rule

$$i^2 = -1 \tag{7}$$

for the imaginary unit, as given in Table 1. He applied for this purpose an observation that Argand had made in 1806, on using what we now call the Argand plane. Denote a complex number with real and imaginary components equal to a and b respectively as a vector $\mathbf{r}$ in the Argand plane (see Fig. 4):

$$\mathbf{r} = a + ib \qquad \Rightarrow \qquad i\mathbf{r} = -b + ia = \mathbf{r}'. \tag{8}$$

(The implication arrow used here means that the material on its left entails that on its right.) When we plot the result of (8), as we do in Fig. 4, we see that under multiplication with i, the vector $\mathbf{r}$ transforms into a vector $\mathbf{r}'$ perpendicular to it (that is, rotated by $\pi/2$). If this operation is repeated

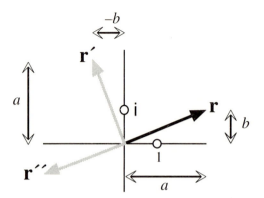

Fig. 4. Multiplications by i and by i^2 are equivalent to counterclockwise rotations by $\pi/2$ and by π respectively.

once more, so that **r′** is multiplied by i and thus rotated again by $\pi/2$ into **r″**, then the original vector **r** changes sign:

$$i\mathbf{r}' = -a - ib = \mathbf{r}'' = -\mathbf{r} \quad \Rightarrow \quad ii\mathbf{r} = -\mathbf{r} \quad \Rightarrow \quad i^2 = -1. \quad (9)$$

Here, on the right of the first implication arrow, we have replaced **r′** by its value given from (8). Clearly, (9) agrees with (7). From this, Hamilton got the idea that the imaginary unit is a rotation by $\pi/2$ or, as he called it, a *quadrantal rotation*. This became a fixed idea in Hamilton's mind, since this was the only way in which he could understand the multiplication rule (7) and this concept, which works perfectly well as long as we keep to the Argand plane, had very sad consequences when applied in other contexts, as we shall see.

An extension of complex numbers: quaternions

Notice how complex algebra already brings about the concept of rotation, which will explain the train of ideas in Hamilton's mind. He had been chasing the possibility of extending the concept of complex numbers to allow for more than one imaginary unit, while he pursued ideas on rotations and on vectors (not yet formally invented but nevertheless already in his mind). He had tried unsuccessfully to define some form of extended complex numbers with two imaginary units. On 16 October 1843, while he was walking to the Irish Academy of Sciences with Lady Hamilton, he had a sudden flash of inspiration and realized that the problem could be solved if three, rather than two, imaginary units were permitted. What is most remarkable is that he immediately saw what the multiplication rules for these units had to be:

$$i^2 = j^2 = k^2 = -1, \tag{10}$$

$$ij = k, \qquad ji = -k. \tag{11}$$

We know this because they were passing Brougham Bridge in Dublin, and there and then Hamilton carved these rules (in a more compact form) with his penknife on the bridge's parapet. A picture of the bridge, and of the plaque that commemorates the now eroded inscription, can be seen in O'Donnell (1983: p. 137).

Notice that (10) is no more than an extension of (7). The real stroke of genius is (11), an equation which has to be extended for the further two cases which arise under the cyclic permutation

$$i \mapsto j \mapsto k \tag{12}$$

(see Table 2). This is the first time in history that a multiplication rule was given which is not commutative, since the product of any two of the imaginary units changes sign whenever the factors are interchanged.

In the same way that the single imaginary unit gives us the complex numbers (1), we now have a super complex number

$$A = a1 + A_x i + A_y j + A_z k, \qquad a, A_x, A_y, A_z \quad \text{real}, \tag{13}$$

which Hamilton, by the time he had arrived at the Academy, already christened with the name of *quaternion*. Just as the complex numbers were written as couples in (4), we can redefine (13) as a couple made up of a scalar a and a vector $\mathbf{A}$ of components A_x, A_y, A_z:

$$A = [a, \mathbf{A}], \qquad \mathbf{A} = (A_x, A_y, A_z). \tag{14}$$

This, of course, will be valid only if the product of two such couples gives us another couple of the same nature, as we shall see in a moment. I must first explain, however, that I am not following here Hamilton's original notation, since he did not even invent the word *vector* until 1846. Even worse, I shall use the modern notation for vectors, introduced by Gibbs in 1881. Although all this will be a bit heretical from Hamilton's point of view, it will allow us to see things far more clearly. Before we do try to multiply quaternions, it is useful to produce a complete multiplication table of the quaternion units, as we do in Table 2, which is set up just like Table 1.

It is now quite easy to multiply two quaternions

$$A = a1 + A_x i + A_y j + A_z k = [a, \mathbf{A}], \tag{15}$$

$$B = b1 + B_x i + B_y j + B_z k = [b, \mathbf{B}]. \tag{16}$$

The result is

$$AB = [ab - \mathbf{A} \cdot \mathbf{B}, a\mathbf{B} + b\mathbf{A} + \mathbf{A} \times \mathbf{B}]. \tag{17}$$

If you are familiar with the rules of vector algebra, it will take you only a

Table 2. Multiplication table of the
quaternion units.

	1	i	j	k
1	1	i	j	k
i	i	−1	k	−j
j	j	−k	−1	i
k	k	j	−i	−1

The product ij appears on the intersection
of the ith row with the jth column.

couple of minutes to obtain (17), but I do not want at this stage to get into
the details of how this is done since we shall go over this work later on in
this chapter (see p. 61). It is sufficient for our present purposes to
remember that $\mathbf{A} \cdot \mathbf{B}$ is a *scalar*, which vanishes whenever the vectors $\mathbf{A}$ and
$\mathbf{B}$ are perpendicular. The so-called cross or vector product $\mathbf{A} \times \mathbf{B}$, instead,
is a *vector*, perpendicular to the plane that contains $\mathbf{A}$ and $\mathbf{B}$. We thus see
that the right-hand side of (17) is indeed a couple of the right form:
everything on the left and right of the comma is a scalar and a vector,
respectively.

We must continue to follow the same pattern of the work for complex
numbers, recognizing that quaternions are extensions of real numbers and
of vectors and we should try to relate quaternions with these objects. As in
(6), real numbers are no trouble:

$$\text{Real number:} \qquad a = [a, \mathbf{0}]. \tag{18}$$

Vectors, on the other hand, are a bit of a problem. The nearest we can get
to them is by defining a *pure quaternion* as a quaternion the scalar part of
which vanishes:

$$\text{Pure quaternion:} \qquad \mathbf{A} = [0, \mathbf{A}]. \tag{19}$$

It is here that Hamilton makes his first mistake. He reasons this way. We
conventionally agree in (18) that a quaternion with a null vector part shall
be called a real number. Likewise, we *define* a quaternion with a null scalar
part to be an object which we shall call a *vector*. Luckily for him, he could
say that, or words to that effect, because vectors did not formally exist until
Hamilton produced this definition:

$$\text{Hamilton:} \qquad \mathbf{A} = [0, \mathbf{A}]. \tag{20}$$

This is so plausible that it appears to be unimpeachable and one is so
used to receiving mathematical verities this way that it is difficult to imagine
that there is anything wrong here. Let us be fair to Hamilton: logically
speaking there was nothing wrong here, and since he was inventing the
word *vector* he was entitled to call this object anything he wanted. The

disaster arose when he went on and tried to use these vectors as we use ours, for instance to denote points in space. For, as we shall see, pure quaternions are never vectors but some very particular rotations.

▌ Quaternions and rotations

As you can see, I am not very good at telling a murder story, since, contrary to the rules of the game, every time there is a clue I raise the reader's attention. We must, however, put off our suspicions for the time being and try to play the game as Hamilton did. It is indeed remarkable that on the same evening of 18 October 1843, when he invented quaternions, Hamilton already knew how they related to rotations. The fundamental property of rotations is that they do not change the length of the vectors on which they are applied. Remember now that vectors were, for Hamilton, a special case of quaternions. So, in order to define rotations he had first of all to define the *length* or *norm* $|A|$ of a quaternion $[a,A]$, which he did as follows:

$$|A|^2 = a^2 + A^2. \tag{21}$$

This is a plausible definition and Hamilton had in fact excellent reasons for it. Once he did this, he defined a quaternion of unit norm as a *normalized quaternion* and it was very easy for him to prove the following result:

The product of two normalized quaternions is a normalized quaternion.

This is a very important result indeed. First of all, it is very easy to produce normalized quaternions, since, for any angle α, they are of the form

$$A = [\cos \alpha, \sin \alpha \, \mathbf{n}], \qquad |\mathbf{n}| = 1. \tag{22}$$

The right-hand side of (21) is, in fact, $\cos^2 \alpha + \sin^2 \alpha \, |\mathbf{n}|^2$, which is unity. The main property of a rotation, from what we have said, is that it effects the following transformation:

$$\text{unit vector } \mathbf{r} \quad \mapsto \quad \text{unit vector } \mathbf{r}'. \tag{23}$$

Hamilton, as we have seen, regards the unit vector $\mathbf{r}$ as identical with the pure quaternion $[0,\mathbf{r}]$, which is a pure normalized quaternion. All that he needs, therefore, in order to obtain a rotation is to transform the pure normalized quaternion $[0,\mathbf{r}]$ into another pure normalized quaternion $[0,\mathbf{r}']$. If we form the product $A[0,\mathbf{r}]$, on using the normalized quaternion in (22), we are automatically sure, from the theorem emphasized in the box above, that the product is a normalized quaternion. All that we now need is to ensure that this quaternion is also pure, that is, that it is a vector within Hamilton's rules, and we have performed a rotation. We are now ready to see how this is done, and to witness one of the greatest cons ever perpetrated in mathematics.

Before we go into this: if any reader thinks that rotations, after all, are geometrical operations, that they are much better studied by geometry and that we are following a crazy way of doing things, then he or she is absolutely right and has my warmest sympathy. We must put ourselves, however, in the spirit of the *Naturphilosophie* of the 1840s and of a Pythagorean mysticism in which numbers and their relations are seen to dominate the universe: such menial tasks as geometrically rotating spheres must be left to the vulgar people (of whom more later).

■ The rectangular transformation

We shall now put into effect the plan sketched and form the product Ar between the two normalized quaternions A from (22) and r, where the latter is the pure quaternion [0,r], with |r| equal to unity. (Remember that for Hamilton this product was a quaternion acting on a vector, just as a rotation does act on a vector!) From (17),

$$\text{Ar} = [\cos \alpha, \sin \alpha \, \mathbf{n}][0,\mathbf{r}] = [0 - \sin \alpha \, \mathbf{n} \cdot \mathbf{r}, \cos \alpha \, \mathbf{r} + \sin \alpha \, (\mathbf{n} \times \mathbf{r})]. \quad (24)$$

We know, from the theorem in the last section, that the quaternion on the right-hand side here must be normalized, as we need, but, alas, it is not a pure quaternion and thus it is not a vector in the Hamilton sense. This, however, is easy to remedy: if we require that $\mathbf{n} \cdot \mathbf{r}$ be null (that is, that $\mathbf{r}$ be perpendicular to $\mathbf{n}$), then

$$\text{Ar} = [0, \cos \alpha \, \mathbf{r} + \sin \alpha \, (\mathbf{n} \times \mathbf{r})] = [0,\mathbf{r}'] = \text{r}', \quad (\mathbf{n} \cdot \mathbf{r} = 0), \quad (25)$$

just as we needed. The vector $\mathbf{r}'$ is

$$\mathbf{r}' = \cos \alpha \, \mathbf{r} + \sin \alpha \, (\mathbf{n} \times \mathbf{r}). \quad (26)$$

We have not yet reached our goal, but we must allow Hamilton to kick the ball into place by identifying the pure quaternions in (25) with their corresponding vectors, on using the rule he gave in (19), a move for which I disclaim all personal responsibility. Equation (25) takes now the form

$$\text{Ar} = \text{r}'. \quad (27)$$

This is precisely a transformation of the form (23) and its geometrical meaning can immediately be recognized from Fig. 5. The vector $\mathbf{r}$ is normal to the vector $\mathbf{n}$, and the vector $\mathbf{n} \times \mathbf{r}$ is perpendicular to $\mathbf{r}$ and to $\mathbf{n}$ and,

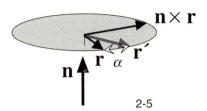

2-5   Fig. 5.   The rectangular rotation.

because **n** and **r** are unit vectors, it is also a unit vector. From (26) the vector **r**′ is in the plane given by **r** and **n** × **r**: it is, like **r**, perpendicular to **n** and it has components cos α and sin α along **n** and **n** × **r** respectively. It must therefore form an angle α with **r**, and thus it is the result of rotating **r** by α about the vector **n**. The result of all this is that Hamilton interprets the quaternion A as a rotation $R(\alpha\mathbf{n})$ by the angle α about the axis **n**. That is:

$$\mathsf{A} = [\cos \alpha, \sin \alpha\, \mathbf{n}] \quad \mapsto \quad R(\alpha\mathbf{n}),$$

α: rotation angle,   **n**: rotation axis   (Hamilton!).   (28)

I must stress that the name of rectangular rotation is my own and not Hamilton's, who referred to this operation just as a rotation. Figure 5, however, is a particular case of the general rotation treated in Fig. 3. For Hamilton, nevertheless, Fig. 5 was the essential thing. The general case of Fig. 3 could be treated by quaternions, but it did not have the nice form that Hamilton liked of a quaternion acting on a vector giving another vector (it turned out, moreover, that the rotation experienced by the vector was by double the angle of the quaternion involved!). Equation (27) became part of the quaternion dogma, and despite being arrant nonsense (remember that it is a transcription of the correct equation (25) and that it depends on Hamilton's identification of pure quaternions with vectors which, as we shall see, is unacceptable) it was taken to be the primary property of quaternions. If, in fact, you look at the second, 1989, edition of the Oxford English Dictionary you will see that a quaternion is defined as an 'operator which changes one vector into another', a definition that comes straight out of equation (27) via Hamilton (1853: p. 122). The next edition, fortunately, will not contain this statement: I am sorry I was too late in preventing its publication in 1989!

I am sure that despite the fact that I have been warning you all the time that the work we have been doing is wrong, it is well nigh impossible at this stage to see why this is so, so subtle being the mirage that Nature set up in front of Hamilton. Why was Hamilton so fooled? For he was one of the sharpest minds of his century, and it is quite clear that the dubious relation between the rotation we have just studied and the one that he called the conical transformation (Fig. 3) was one that he did not take lightly. There was, however, another false trail that Nature had set up to persuade Hamilton that his approach, as summarized in (28), was right. Let us take the angle of rotation α in this equation to be $\frac{\pi}{2}$:

$$[\cos \tfrac{\pi}{2}, \sin \tfrac{\pi}{2}\,\mathbf{i}] = [0, \mathbf{i}] = \mathrm{i}. \tag{29}$$

Therefore, in the sense that for Hamilton a quaternion such as **A** in (28) is a rotation by α, then the imaginary units are rotations by $\tfrac{\pi}{2}$ or quadrantal rotations,

$$\mathrm{i} \quad \mapsto \quad R(\tfrac{\pi}{2}\mathbf{i}), \tag{30}$$

just as required by the Argand construction in Fig. 4 and as (seemingly!) necessary in order to understand why the square of the imaginary units is -1. Everything thus fitted the facts perfectly. And yet, as we shall soon see, nothing was right. That this was the case was the result of some work done in Paris, in fact a little before Hamilton's discovery of the quaternions, by an obscure (to the British) socialist banker, Olinde Rodrigues, whose story we must now tell.

▌Olinde Rodrigues

You could not imagine a person more different from Hamilton in background and outlook than Olinde Rodrigues. He was born in Bordeaux on 6 October 1795, the son of a Jewish accountant belonging to an old established banking family of Spanish origin. So little is known about Rodrigues that until a year or two ago even his date of birth was wrongly quoted. We do not even know how he was educated and how he became a mathematician, except that in 1816 he was awarded a doctorate by the then newly established University of Paris. We know that he could not have attended the École Polytechnique because of his Jewish origin, and, whereas there was some idea that he might have attended the École Normale, it is now almost certain that he did not. After 1816 he is lost to the world of mathematics, in fact to the world at large, because we know practically nothing about him until May 1823. We know that he did at first some freelance teaching for a pupil at the École Polytechnique and that he then engaged himself in banking in Paris. By 1823 he must have been doing pretty well financially because he was then introduced by a common friend to a great man in severe need of assistance as one likely to be able to provide it. This man was the charismatic utopian socialist Saint Simon, who, having failed to blow his brains with three pistol shots a couple of months before, was lying ill and destitute. Since then, we know Rodrigues as a social reformer and as the intellectual and political successor of Saint Simon. He was also the director of a bank specializing in mortgages for housing, and he wrote extensively on financing of social projects, and on labour legislation. In his later life, he took a part in the development of the French railways, thus starting a tradition that was to remain influential in

France up to our time. Of mathematics, except for the little teaching mentioned, we know nothing about any work or even of any interest he may have had since the time of his thesis. It is abundantly clear, however, that he was intellectually as far away from *Naturphilosophie* as any man of his time. As a socialist and as a French intellectual brought up under the spirit of the enlightenment and the rationalism of Diderot's *Encyclopaedia*, there is no doubt that Hamilton's task of interpreting 'the oracles of the universe' could not have been more alien to him.

All this is very clearly reflected in the one mathematical paper that out of the blue Rodrigues published in 1840 (Rodrigues 1840), almost twenty-five years after appearing to have abandoned mathematics, a subject to which he never returned again for the rest of his life. This paper is concerned with the study of rotations and in it, three years before Hamilton invented quaternions, he had all the answers to the problems that bedevilled Hamilton. Not only is Rodrigues' mathematical approach diametrically opposite to that of Hamilton, but his writing style is just from another planet. Whereas Hamilton invents dozens of terms just to describe the simplest of concepts, and thus achieves an almost unmatched obscurity in his writing, Rodrigues starts by assuming that he is dealing with a simple problem for which he has to produce the simplest possible answer. He has not got to look over his shoulder all the time, like Hamilton did, to try to fit his results within a universal scheme of things. The best way in which we can appreciate this is by looking at Rodrigues' work and comparing his results with those that Hamilton was to produce three years later.

∎ Rodrigues and rotations

Euler had proved in 1775 that a rotation by β about the axis $\mathbf{m}$, $R(\beta\mathbf{m})$, followed by a rotation by α about $\mathbf{l}$, $R(\alpha\mathbf{l})$, must equal a rotation by some angle γ about some new axis $\mathbf{n}$:

$$R(\alpha\mathbf{l})\,R(\beta\mathbf{m}) = R(\gamma\mathbf{n}). \tag{31}$$

Notice that on the left-hand side of (31) the first operation listed is the second one performed, as is the case when, say, we write log sin, meaning that the sin is taken first and then the log. The first problem that Rodrigues faced was to find the precise geometrical relation between these three rotations, so as to be able, given the left-hand side of (31), to obtain the axis and angle of rotation of the resultant operation on the right-hand side. This he did by a very ingenious but simple geometrical construction (unfairly called to this day the *Euler construction*). The result of this construction is that the six geometrical elements in (31) are related by the spherical triangle shown in grey in Fig. 6.

In order to understand this figure we must appreciate that, for example, the first operation performed on the left of (31), $R(\beta\mathbf{m})$, is best understood, as all rotations are, by means of its action on a sphere (conveniently taken to be of unit radius), the origin O of which coincides with the point of application of the *rotation axis* $\mathbf{m}$. Thus $R(\beta\mathbf{m})$ rotates this sphere about $\mathbf{m}$ by the angle β, and this statement fully defines the rotation in question.

Figure 6 deals of course with a far harder problem, which is the realization of equation (31). What that equation says is that if we first rotate the sphere about the axis $\mathbf{m}$ by the angle β and we then rotate the sphere in its resulting position now about the axis $\mathbf{l}$ by the angle α, then the sphere would be left in precisely the same orientation as if it had been rotated just once only, about the axis $\mathbf{n}$ by the angle γ. The great merit of Rodrigues' construction is that, given $\mathbf{l}$, $\mathbf{m}$, α, and β it provides the axis $\mathbf{n}$ and the angle γ of the single resulting rotation.

It is not necessary for us to give the theory behind Rodrigues' construction in Fig. 6 (see, for example, Altmann 1986: p. 155), but it is useful to understand its result, that is, the prescription which it provides, which is

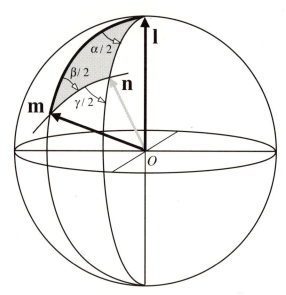

Fig. 6.   The product of the rotation by β about $\mathbf{m}$ followed by the rotation by α about $\mathbf{l}$ equals the rotation by γ about the new axis $\mathbf{n}$. In order to make the reading of this picture easier, the vector $\mathbf{l}$ has been taken at the 'north pole' of the sphere, but its orientation can be entirely general. It can be easily seen in this way that the curve that joins the heads of $\mathbf{l}$ and $\mathbf{m}$ is a great circle of the sphere. This great circle is rotated counterclockwise about $\mathbf{l}$ by $\frac{\alpha}{2}$ and clockwise about $\mathbf{m}$ by $\frac{\beta}{2}$. Thus the curve shown joining the heads of $\mathbf{m}$ and $\mathbf{n}$ is also part of a great circle, having being obtained by rotation of the great circle part of which is in thick line.

as follows: given the axes **l** and **m** of the rotations being multiplied, one side of the spherical triangle (shown highlighted in the figure) is determined by taking the arc of the great circle that passes through the heads of those two axes. If we then rotate this great circle clockwise for the first rotation (about **m**) and counterclockwise for the second, in each case by *half* the corresponding rotation angle, then the position of the resulting axis **n** is determined by the point of intersection of the two arcs that are thus formed. The angle γ of the resulting rotation is at the same time determined as follows: consider the spherical triangle shown in grey in the figure; half the angle of rotation, $\frac{\gamma}{2}$, is the supplementary angle shown in the figure.

Having demonstrated how this prescription arises, Rodrigues then proceeded to solve the spherical triangle that he had obtained, in order to get the angle and axis of the resultant rotation, γ and **n**, in terms of those of the given rotations. I quote below the results which Rodrigues obtained, in the same way in which he presented them, except that I use the modern vector notation (Rodrigues' formulae contained all the components explicitly):

$$\cos \tfrac{\gamma}{2} = \cos \tfrac{\alpha}{2} \, \cos \tfrac{\beta}{2} - \sin \tfrac{\alpha}{2} \, \sin \tfrac{\beta}{2} \, \mathbf{l} \cdot \mathbf{m}, \qquad (32)$$

$$\sin \tfrac{\gamma}{2} \, \mathbf{n} = \sin \tfrac{\alpha}{2} \, \cos \tfrac{\beta}{2} \, \mathbf{l} + \cos \tfrac{\alpha}{2} \, \sin \tfrac{\beta}{2} \, \mathbf{m} + \sin \tfrac{\alpha}{2} \, \sin \tfrac{\beta}{2} \, \mathbf{l} \times \mathbf{m}. \qquad (33)$$

Although these equations were published three years before Hamilton's discovery of quaternions, they entail the quaternion multiplication rule. It is in fact sufficient to associate with each rotation a quaternion of the following form

$$R(\alpha \mathbf{l}) = [\cos \tfrac{\alpha}{2}, \, \sin \tfrac{\alpha}{2} \, \mathbf{l}], \qquad R(\beta \mathbf{m}) = [\cos \tfrac{\beta}{2}, \, \sin \tfrac{\beta}{2} \, \mathbf{m}], \qquad (34)$$

and similarly for $R(\gamma \mathbf{n})$. It is then obvious that the quaternion multiplication rule (17) applied to the product of the two quaternions in (34) gives the quaternion for $R(\gamma \mathbf{n})$ exactly as in (32) and (33):

$$[\cos \tfrac{\alpha}{2}, \, \sin \tfrac{\alpha}{2} \, \mathbf{l}][\cos \tfrac{\beta}{2}, \, \sin \tfrac{\beta}{2} \, \mathbf{m}] = [\cos \tfrac{\gamma}{2}, \, \sin \tfrac{\gamma}{2} \, \mathbf{n}]. \qquad (35)$$

The remarkable result of this work is that whereas Hamilton, as shown in (28), associated with $R(\alpha \mathbf{l})$ a quaternion $[\cos \alpha, \, \sin \alpha \, \mathbf{l}]$ the angle of rotation now appears in the quaternion as a half-angle. This has extraordinary consequences and it is the great achievement of Rodrigues that it was he who for the first time introduced half-angles in the study of rotations. For people of our generation this appears to be an event just like any other. For Hamilton, as we shall explain below, it was something that would have deeply offended his view of the universe. Before we go into this, however, notice that what we are saying is that *the angle of rotation is double the angle which appears in the quaternion*. For Hamilton, on the other hand, the latter was the rotation angle. If we go back to Fig. 3 this explains why, for Hamilton, the angle α in that picture was *twice* the rotation angle!

The rectangular transformation unmasked

Before we go into the reasons why Hamilton could never have accepted Rodrigues' parametrization of rotations through half-angles, let us now try and make sense of the rectangular transformation from Rodrigues' point of view, since this transformation appears so far to be a simple and conclusive corroboration of Hamilton's ideas. Let us examine, first of all, what is the meaning of a pure normalized quaternion $[0,\mathbf{r}]$, with $|\mathbf{r}|$ unity. The kingpin of Hamilton's argument was that this quaternion could be identified with the vector $\mathbf{r}$ (see equation (20)). Let us write this quaternion in the form (34):

$$\mathbf{r} = [0,\mathbf{r}] \equiv [\cos \tfrac{\pi}{2}, \ \sin \tfrac{\pi}{2}\mathbf{r}] = R(\pi\mathbf{r}). \tag{36}$$

What we discover now is that a pure quaternion cannot be identified at all with a vector: it is merely a rotation by π, that is, what one calls a *binary rotation*. As you will remember, what we had obtained in equation (25) is

$$\mathbf{Ar} = \mathbf{r}'; \qquad \mathbf{A} = [\cos \alpha, \sin \alpha\, \mathbf{n}], \tag{37}$$

with

$$\mathbf{r} = [0,\mathbf{r}], \quad \mathbf{r}' = [0,\mathbf{r}']; \qquad \mathbf{r} \perp \mathbf{n}, \quad \mathbf{r}' \perp \mathbf{n}, \quad \angle(\mathbf{r},\mathbf{r}') = \alpha. \tag{38}$$

The expression $\mathbf{Ar} = \mathbf{r}'$ must now be understood from (35), as quaternion products are, as a product of two rotations giving another rotation. More specifically, with the identification of $\mathbf{A}$ from (37) and of $\mathbf{r}$ and $\mathbf{r}'$ from (36), it states that the given rotation by 2α times a binary rotation about $\mathbf{r}$ gives a binary rotation about $\mathbf{r}'$. Hamilton, instead, read this result as

$$\mathbf{Ar} = \mathbf{r}', \qquad \mathbf{r} = [0,\mathbf{r}], \qquad \mathbf{r}' = [0,\mathbf{r}'], \tag{39}$$

which for him it meant 'the quaternion $\mathbf{A}$ acting on the vector $\mathbf{r}$ transforms it into the vector $\mathbf{r}'$'. The true meaning of (37), however, is very simple indeed, if we identify all the quaternions in it with the corresponding rotations, now in full detail:

> The product of a rotation by 2α about the axis $\mathbf{n}$, times a binary rotation about the axis $\mathbf{r}$, $(\mathbf{r} \perp \mathbf{n})$ is another binary rotation about the axis $\mathbf{r}'$, $(\mathbf{r}' \perp \mathbf{n})$ at an angle α from $\mathbf{r}$.

That is:

$$R(2\alpha\mathbf{n})\, R(\pi\mathbf{r}) = R(\pi\mathbf{r}'); \qquad \mathbf{r} \perp \mathbf{n}, \quad \mathbf{r}' \perp \mathbf{n}, \quad \angle(\mathbf{r},\mathbf{r}') = \alpha. \tag{40}$$

Anyone with the slightest knowledge of crystallography will recognize this relation as one of the corner-stones of geometrical crystallography: if there is a rotation axis **n** by 2α perpendicular to a binary axis (**r**) then there must be another binary axis (**r′**) also perpendicular to **n** and at an angle α with **r**. This is precisely the relation between the axes that we see in Fig. 5. (Remember that, from equation (24), the angle of the rotation about the axis **n** is in the correct Rodrigues parametrization 2α.)

So as not to leave the slightest possible doubt on this point, let us multiply the two operations on the left of (40) geometrically, which we do in Fig. 7. This figure is to be understood as follows. In order to perform rotations we must follow what happens to one point of a sphere when the sphere is rotated. The circle in the figure is the equatorial plane of the sphere assumed symmetrically disposed above and below the plane of the figure. In the product on the left of (40) we must take the operations from right to left, that is, we must start with the binary rotation. The point 1 above the equator is taken under this rotation by π about **r** to a diametrically opposed point 2 below the equator. The rotation by 2α about the axis **n** normal to **r** takes the point 2 to the point 3. It can be seen at once that the original point 1 can be taken to the final point 3 by a binary rotation about **r′**. (Notice in fact that in this rotation the point 1 above the plane and at a certain angle, say β, with **r′** must go into another point below the plane and at the same angle β with **r′**, which is indeed the point 3.) This operation is therefore the product sought, which agrees with the right-hand side of (40).

On comparing this figure with Fig. 5, we can now understand perfectly well the optical illusion that Hamilton suffered: he read **r** and **r′** in Fig. 5 as icons of vectors and not as icons of rotation axes, as they clearly are in Fig. 7. He thus thought that the quaternion A transformed the vector **r** into the vector **r′** ('rotated' from **r** by α) instead of recognizing that the quaternion A multiplied by the quaternion of a binary rotation gives the

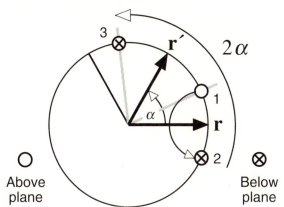

Above plane

Below plane

Fig. 7. Product of a binary rotation about **r** followed by a rotation by 2α about an axis **n** normal to the plane of the figure. The angle β mentioned in the text, between the grey line and the axis **r′** has been left implicit for clarity.

quaternion of a second binary rotation and that the two binary rotation axes are not obtained one from the other by any rotation at all. It is understandable that Hamilton read this situation in this way, because he had serious moral reasons, as we shall see, to try and defend at all costs his rotation parametrization. What is amazing is that whereas Hamilton's works have been analysed with greater intensity than those of perhaps any other 19th century mathematician, this mistake was not understood for well over a century! Hamilton's moral problem as regards the parametrization of rotations will now be grasped when we look at what happens to the imaginary units from the point of view of Rodrigues.

∎ The quaternion units

We have already seen in (36) that a pure quaternion can never be a vector and that it always is a binary rotation. It follows that this must also be the case, in exactly the same way, for the imaginary units:

$$r = [0,\mathbf{r}] = R(\pi\mathbf{r}), \tag{41}$$

$$i = [0,\mathbf{i}] = R(\pi\mathbf{i}). \tag{42}$$

Remember that, as stated in (19), Hamilton made no notational distinction between a pure quaternion and its corresponding vector, and thus, for him, the quaternion unit i and the vector $\mathbf{i}$ were one and the same thing. Equation (42), instead, tells us that the quaternion unit i is a binary rotation about the vector $\mathbf{i}$, rather than being identical to the latter. We can thus see how injudicious it was not to make the distinction between the quaternion i and the vector $\mathbf{i}$, specially if one recognizes that Hamilton's notation was so successful that it is still used in our own day for the famous unit vectors $\mathbf{i}$, $\mathbf{j}$, and $\mathbf{k}$. It is rather disturbing to realize that such important vectors started life in this preposterous way. You must remember that for Hamilton the imaginary units were quadrantal rotations, an essential point for him to be able to understand how their square could equal -1, as we have seen in Fig. 4 and in (30). We can now see very well how the identification of the imaginary units with binary rotations would have been entirely unacceptable for Hamilton. If an imaginary unit is a rotation by π, its square must be a rotation by 2π, so that it must now be the rotation by 2π that multiplies its operand by -1:

$$i^2 = -1 \quad \Rightarrow \quad R(2\pi\mathbf{i}) \quad \mapsto \quad -1 \tag{43}$$

(compare with equation (9)).

Whereas such behaviour was crystal-clear for a rotation by π, the present description is incomprehensible, if not heretical, since the rotation by 2π is no rotation at all: it is the identity, and as such it should not change the sign of anything. It is child's play to recognize in (34) that, when the rotation angle is 2π a change of sign is entailed with respect to a rotation by nothing, so that it is most unlikely that Rodrigues was not aware of this strange result. Admittedly, he says nothing about it in his paper. He did not carry, like Hamilton, the responsibility for the whole world on his shoulders: he knew that he had done a simple, geometrical, rational job and that the answer had to be right. I believe, on the other hand, that Hamilton was quite aware of this possibility, for he was undoubtedly mystified by the strange fact that in Fig. 3 he had to take α as *twice* his angle of rotation. We can surmise this, because when, through a rather subtle quaternion procedure, he managed to obtain this result, he did not publish it, contrary to his habit, and it was Cayley who took the priority in print for what was called the *conical transformation*. I am sure that this problem went through Hamilton's mind many times but the possibility that the identity rotation (realized as a rotation by 2π) could be required to change the sign of anything was one that he could not countenance. He could not either, of course, foresee the answer to this conundrum which took sixty years or so after the discovery of quaternions. This will be our last piece in this jigsaw puzzle, but in order to understand the ideas that had to be invented, we must go back to vectors.

▌ Vectors again

We saw in equation (17) how quaternions multiply and, in doing so, we had to accept for the time being the use of some vector multiplication rules. The appearance of these rules was one of the most important by-products of Hamilton's algebra and we shall see that it led to a most important idea that allowed Cartan many years after Hamilton to discover some remarkable geometrical objects which do change sign after a rotation by 2π.

Given two quaternions,

$$A = a1 + A_x i + A_y j + A_z k = [a,\mathbf{A}], \tag{44}$$

$$B = b1 + B_x i + B_y j + B_z k = [b,\mathbf{B}], \tag{45}$$

we have written their product in (17) as

$$AB = [ab - \mathbf{A} \cdot \mathbf{B}, a\mathbf{B} + b\mathbf{A} + \mathbf{A} \times \mathbf{B}]. \tag{46}$$

It will now be useful to go in some detail into how this result is obtained. The idea is very simple indeed: all we have to do is to multiply the two four-term expressions in (44) and (45) and whenever a product such as $A_x B_y ij$ appears, we look up Table 2 (p. 50) and find the value of the product of quaternion units that we have. It will be convenient to give here again for ease of reference the values of these products:

$$i^2 = j^2 = k^2 = -1;$$ (47)

$$ij = k, \quad ji = -k; \qquad jk = i, \quad kj = -i; \qquad ki = j, \quad ik = -j. \quad (48)$$

What we are interested in are the nine terms in the product $\mathbf{AB}$ of (44) and (45) that contain two quaternion units. Three of them will be of the following form:

$$A_x B_x i^2 + A_y B_y j^2 + A_z B_z k^2 = -(A_x B_x + A_y B_y + A_z B_z) = -\mathbf{A} \cdot \mathbf{B}. \quad (49)$$

The so-called scalar product of two vectors, $\mathbf{A} \cdot \mathbf{B}$, is defined in fact, subject to some simple conditions, as the sum of the products of components in the bracket in (49) as follows. If

$$\mathbf{A} = (A_x, A_y, A_z), \qquad\qquad \mathbf{B} = (B_x, B_y, B_z), \quad (50)$$

then

$$\mathbf{A} \cdot \mathbf{B} = A_x B_x + A_y B_y + A_z B_z. \quad (51)$$

The remaining six terms will separate out in three pairs of the form

$$A_x B_y ij + A_y B_x ji = (A_x B_y - A_y B_x)k, \quad (52)$$

where we use the products in (48). Those familiar with vector algebra will recognize that the term in brackets here is the z component (that is, the one corresponding to $\mathbf{k}$) of the vector product $\mathbf{A} \times \mathbf{B}$. The other four terms remaining make up in the same manner the x and y components of this product. Let us have a quick look at the properties of this vector product, which we shall equate to a vector $\mathbf{C}$:

$$\mathbf{C} = \mathbf{A} \times \mathbf{B} = (C_x, C_y, C_z) = (A_y B_z - A_z B_y, A_z B_x - A_x B_z, A_x B_y - A_y B_x). \quad (53)$$

I must now come clean. I have called this object a vector but in fact I do not have a right to do so. The mere fact that it has three components is no guarantee that it is a vector. Let me sketch how we would have to proceed in order to licence $\mathbf{C}$ as a vector. First, we would have to establish very precise transformation rules of the vectors $\mathbf{A}$ and $\mathbf{B}$ in (50) under all rotations and reflections. (We would have to accept for this purpose that $\mathbf{A}$ and $\mathbf{B}$ are ordinary, polar, vectors.) Once this is done we would have to use some algebra to find out how the three components in (53) behave: if they behave exactly as their parent vectors, then they are a vector. Remember in understanding this that an important principle of modern life is that what gives a geometrical entity such as a vector its name is the way in which it transforms. This work, alas, is far too complicated for us, so that you will have to accept that (53) passes the rotation test with flying colours: it behaves exactly like a vector in this respect.

Fortunately, it is a lot easier to see how (53) transforms under reflection. We take for this purpose a particularly simple case, in which both A_z and B_z vanish, whence the (for the time being alleged) vector $\mathbf{C}$ has only one component, which makes the geometry much simpler:

$$\mathbf{C} = (0,0,A_xB_y - A_yB_x). \tag{54}$$

We want to study the way in which this vector behaves under reflection through the plane shown in Fig. 8. We must be extremely careful not to jump to conclusions in doing this. What we mean by the statement just given is this: we know that $\mathbf{C}$ is a product of two vectors $\mathbf{A}$ and $\mathbf{B}$ and we assume that we know how these vectors perform under reflection, for which purpose we must take them to be ordinary or polar vectors, in the terminology of our first chapter. We must induce the reflection of $\mathbf{C}$ (a vector the behaviour of which cannot yet be presumed known) by reflecting the known vectors $\mathbf{A}$ and $\mathbf{B}$. Now, for any vector $\mathbf{V}$ (see Fig. 8),

$$\sigma V_x = V_x, \qquad \sigma V_y = -V_y. \tag{55}$$

Therefore,

$$\sigma\mathbf{C} = \sigma(0,0,A_xB_y - A_yB_x) = (0,0,-A_xB_y + A_yB_x) = -\mathbf{C}. \tag{56}$$

We must realize that $\mathbf{C}$ has only a z component and it is thus parallel to the reflection plane σ, and that on reflection through this plane it performs the infamous trick of changing sign, which is the genetic fingerprint of axial vectors. We have thus proved the important result that *the cross product of two polar vectors is not a polar but an axial vector.*

What we have learnt in this section is that by taking some products of vectors, that is, by forming objects the components of which are products of vector components, we can generate new objects. The second important principle is that, however unfamiliar their expressions be, one can recognize what these objects are as long as one can identify their transformation properties: any object with three components that transforms exactly like a vector *is* a vector!

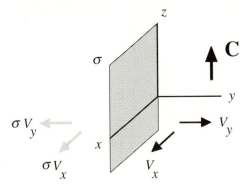

Fig. 8. A reflection plane perpendicular to the y axis.

The next thing we shall now do is to summarize what we have done in order to devise some machinery which will allow us to construct objects with the magic property that they change sign under rotations by 2π, and which are called *spinors*.

■ Tensors and spinors

What we have done in the last section is basically this. Given two vectors,

$$\mathbf{A} = (A_x, A_y, A_z), \qquad \mathbf{B} = (B_x, B_y, B_z), \tag{57}$$

we can imagine that we construct a set of nine components,

$$A_x B_x, A_x B_y, A_x B_z, A_y B_x, A_y B_y, A_y B_z, A_z B_x, A_z B_y, A_z B_z. \tag{58}$$

The object so defined (that is, the set in equation (58)) is called a *tensor*, in fact of *rank two*, because each element of it is a product of two components. Correspondingly, the operation by which this tensor is obtained from the two vectors $\mathbf{A}$ and $\mathbf{B}$ is called a *tensor product*. Out of this tensor we have seen that we can separate two distinct objects, one a scalar, formed as in (51) by the *symmetric* components, (that is, by the components that do not change sign when the suffices are exchanged),

$$A_x B_x + A_y B_y + A_z B_z, \tag{59}$$

and another object of three *antisymmetric* components, like in (53),

$$A_y B_z - A_z B_y, \quad A_z B_x - A_x B_z, \quad A_x B_y - A_y B_x. \tag{60}$$

(Notice that, in fact, the first component does change in sign when the suffices y and z are exchanged, and similarly for the other components.)

Because this latter object has three components, it can masquerade as a vector and it can in fact be proved that it transforms under all rotations exactly as a vector does. This is what allows this object to impersonate a vector so successfully that one calls it a *vector*, or, more properly, an *axial vector*. Where one can see that it is not quite a vector is through the fact that it transforms pathologically under reflections, as we have shown in (56). Mathematically, its proper name is *antisymmetric tensor of rank two*. The important thing that you must remember is that axial vectors turn up when forming tensor products of ordinary (polar) vectors.

We must be grateful to Hamilton for having given to us the germ of these important ideas, from which we must now move on. We want to see whether it is possible to conceive of objects which transform in the bizarre manner required by (43) under rotations by 2π, but which are nevertheless compatible with the fact that honest vectors must be left invariant under this rotation, since otherwise we would be falling into nonsense. In order to dream up such objects, we can ask ourselves the following question. We have seen that axial vectors are tensor products masquerading as vectors. Would it be possible that ordinary, that is, polar, vectors are themselves tensor products of some simpler primary objects, these tensor products being so cunningly contrived that they masquerade for our real, everyday, polar vectors? The answer to this question was provided in the affirmative by Cartan in 1913, and it is very simple to grasp, at least if we leave out technical details.

We must first of all dream up objects simpler than vectors: they must be of dimension 1 or 2. If they have only one component, we shall never get three by forming a tensor product, so that we must try two components, say

$$\lambda = (\lambda_1, \lambda_2), \qquad \mu = (\mu_1, \mu_2). \qquad (61)$$

We form their tensor product, as in (58), and we now have four components:

$$\lambda_1\mu_1, \quad \lambda_1\mu_2, \quad \lambda_2\mu_1, \quad \lambda_2\mu_2. \qquad (62)$$

Out of this number we can form one object with three symmetric components,

$$(\lambda_1\mu_1, \lambda_1\mu_2 + \lambda_2\mu_1, \lambda_2\mu_2), \qquad (63)$$

and another object with a single antisymmetric component,

$$\lambda_1\mu_2 - \lambda_2\mu_1. \qquad (64)$$

Because (63) has precisely three components, it is quite possible that it will behave like a three-dimensional vector, and, in fact, it can be proved that under all rotations and all reflections it behaves precisely as a polar vector. Let me explain what I mean by this statement. In order to obtain the transformation properties of (63) under rotations and reflections, one must first guess (mathematicians say 'postulate') transformation rules under these operations for the objects (λ_1, λ_2) and (μ_1, μ_2) in (61). (These rules, of course, must be the same for both objects.) Given these transformation rules it is fairly easy to obtain those for the object in (63), and if the original guess is good then these transformation rules turn out to be identical with the transformation rules for vectors. Although this work is a lot easier than it sounds, it is too technical for me to treat here and the interested reader should consult Altmann (1986: pp. 100–8).

When dealing with objects such as vectors and tensors, we have learnt that what makes an object what it is, is its symmetry behaviour. The result of the work just sketched is, therefore, that polar vectors are given by (63) and that, contrary to what our intuition dictates, they are not simple objects but rather tensorial products of some simpler, two-dimensional objects, given in (61), which are called *spinors*. Let us now assume that it is the spinors that change sign under a 2π rotation, as required by (43), a fact that may be corroborated directly from their transformation rules under rotations (which, as we have said, must be known through postulation). We can then see at once that this property still leaves vectors enjoying a good and peaceful life: if each component λ_i and μ_i in (63) changes sign under a 2π rotation, the three vector components in that equation remain yet invariant! We have now reconciled equation (43), which applies to spinors, with the requirement that the rotation by 2π be the true identity for vectors.

It should be noted that we have invented spinors (or rather Cartan did) from purely geometrical considerations concerning rotations. More than ten years after Cartan, Pauli, when writing down the wave equation for a spinning electron, re-invented the objects defined in (61) with precisely the transformation properties under rotations that had been required by Cartan. This is the way in which these objects came to be called *spinors*, since they describe the electron spin. The Hamilton–Rodrigues story thus closes with the same theme with which Ørsted's ended: the Dane's timid dream of some 'secret' rotations that could explain magnetism (see p. 33).

What a strange story symmetry has unfolded for us. We have found in our first chapter that not all vectors belong to the same species, and we now come to the shocking fact that the everyday vectors that denote points of our everyday space are not simple objects but rather (tensor) products of even more elementary, two-dimensional objects, called spinors. Most people who have heard the word *spinor* think that it is an object which appears only in quantum mechanics, whereas what we now realize is that the spinors characterize a fundamental property of ordinary space. Because this property eluded Hamilton, he could not face the consequences of his own results, whereas it took the cooler, more detached banker Olinde Rodrigues to start the train of thought that brought us to a deeper under-standing of 3-space. It is perhaps not entirely unrelated that as a utopian socialist he was not unhappy to see idols and icons demolished and that he could thus calmly accept in his own mind the rather bizarre requirement that rotations by 2π be associated with a change of sign.

Having thus justified the status of the rotation by 2π, we can happily accept that imaginary units can be realized by binary rotations, since their square (which is a rotation by 2π) gives the necessary factor of -1. I have no doubt that it was Hamilton's vision of those units as quadrantal rotations that guided him in his path through the understanding of rotations with such awkward results. (It must be stressed that what appears as perfectly correct in the Argand plane in Fig. 4 is not necessarily true in ordinary three-dimensional space!) My readers must be warned that this delusion of Hamilton is infectious and that to this very day people write whole chapters in books in order to demonstrate the (untrue) behaviour of the imaginary units as quadrantal rotations!

▌ Epilogue

Hamilton was by our standards a young 38 when he discovered quaternions in 1843. He had already been doing work of international standing for some 22 years and time was taking its toll on him. His next 22 years until his death were not happy ones. People reckon that he had some financial problems. Then, although a respectable family man, he also experienced some amorous anxieties. Some said that he drank too much. I believe that poor Hamilton was aware of the internal contradictions of his doctrine on quaternions and that it was the greatest of misfortunes that, instead of openly ventilating them he, and his school, became dogmatic, inward looking, and intolerant. Hamilton almost certainly never read Rodrigues' paper, although we know that Cayley did so, and Cayley himself was somewhat perplexed by it.

Rodrigues lived for seven or eight years after the invention of quaternions (even the date of his death is not known certainly, being quoted as either 1850 or 1851). That Rodrigues might have heard of quaternions is most unlikely: his last few years were mainly concerned with railway financing and social legislation. He died largely forgotten, his paper on rotations hardly read at all, and the profound connection between his work and the quaternion algebra almost entirely ignored.

The last year of Hamilton's life, 1865, was in contrast a good one for him: three months before his death he was named by the American Academy of Sciences as the top living scientist in their roll of foreign associate members. After this, however, quaternions had a rough time. It is rather curious that even their objectors could not put their fingers precisely on the neuralgic points of the quaternion dogma, and I think that this was partly due to the fact that Hamilton's writings were so obscure.

Although Grassmann and Heaviside were influential, it was mainly Willard Gibbs from Yale who finally tidied up the subject and rescued from Hamilton's work the useful concepts of scalar and vector products, which Hamilton had buried in the morass of his quaternion algebra. In 1881, Gibbs published privately *The Elements of Vector Analysis*, which was the most influential work in achieving the replacement of quaternions by vectors. In doing so, the wonderful opportunity opened up by Rodrigues for using quaternions for their proper object in the study of rotations was entirely neglected. Because the claim had been made for quaternions that they held the key to a super form of geometry, and because that claim had failed, quaternions became suspect even where they could have been wonderfully useful.

The invention of the vector calculus was an effrontery to the quaternionists, who reacted in a dogmatic and almost irrational manner. Thus P. G. Tait, the most influential of the quaternionists, writes in the preface of his treatise on quaternions (Tait 1890: p. vi):

> *Even Prof. Willard Gibbs must be ranked as one of the retarders of quaternion progress, in virtue of his pamphlet on Vector Analysis, a sort of hermaphrodite monster, compounded of the notations of Hamilton and of Grassmann.*

It is somewhat amusing that only two years later Lord Kelvin, who as Sir William Thomson had been Tait's collaborator in their famous *Treatise of Natural Philosophy*, took quite the opposite view (see Thompson 1910: Vol. II, p. 1070):

> *Quaternions came from Hamilton after his really good work had been done; and, though beautifully ingenious, have been an unmixed evil to those who have touched them in any way, including Clerk Maxwell.*

Very often people ask the question: how is it that mathematics, that appears to be the product of pure thought, is so well adapted to describe nature? If people ask this question it is because they have never been exposed to mathematics in the making, which, as we have seen with Hamilton's example, *can go wrong*: it is only by trial and error that one arrives at useful mathematics. It is perfectly feasible, even for a man of Hamilton's genius, to propose conventions that are entirely inadequate, like his identification of pure quaternions as vectors. The dream of the visionary, like Hamilton, has to be tested by the insight of the empirical practical man, like Rodrigues, who offered no conventions, no rules, no grand ideas, but good solid geometrical facts. In mathematics, like in everything else, it is the Darwinian struggle for life of ideas that leads to the survival of the concepts which we actually use, often believed to have come to us fully armed with goodness from some mysterious Platonic repository of truths.

If, by saying this, I seemingly relegate all intellectual rubbish to the dustbin to which it belongs, let me put in a plea in its favour, because what appears to be an entirely unfounded belief often provides a useful drive from which important results emerge. Dreams and fantasies have thus been almost as influential in our intellectual development as the discovery of new facts. It is very tempting, for instance, to dismiss *Naturphilosophie* out of hand as intellectual rubbish. The concept of the unity of Nature, however, led Ørsted to one of the most influential discoveries of the physics of the 19th century, and his Pythagorean belief in the deep significance of numbers led Hamilton to the discovery of a whole new view of algebra which turned out to be most important in modern physics. Right in the middle of the age of the enlightenment, the achievements of these two supporters of *Naturphilosophie* will remain outstanding for all time.

Hamilton's deep and not always successful struggle with the oracles of the universe was not in vain: it produced the line of thought from which vectors, tensors, and spinors were born. And all this comes from the study of that very simple symmetry operation called rotation.

3 Peierls and symmetry breaking

Est modus in rebus, sunt certi denique fines,
Quos ultra citraque nequit consistere rectum.

There is a measure in things, there are certain fixed limits,
beyond and short of which rightness breaks down.

Horace, *Satyres*, I, i, 106

There are two quite separate jobs that I want to do in this chapter. So far, we have talked a fair amount about symmetry operations, such as reflections and rotations, and we probably have got some idea that they are important and even beautiful. We have also seen that they can lead us into interesting theoretical ideas and that, through the correct use of the symmetry principle, we can interpret and even predict some experimental results. We must admit, however, that we have not yet seen that the study of symmetry can do anything more practical for us. We are going to remedy this, because we shall soon show that symmetry operations allow us to classify the energy levels in systems such as crystals and molecules, and we shall see that this classification determines in a very simple way some pretty formidable properties of those systems.

On another different line, we have talked a great deal about the fundamental property of symmetry conservation and we want to look more carefully into what this entails and how this works in realistic cases. For it is obvious that when dealing with real rather than ideal systems, the principle of conservation of symmetry has to be taken with a salubrious pinch of salt. No one, for instance, will expect the cone in Fig. 1 to stay forever in the symmetrical configuration shown: sooner or later it will fall down. We can of course talk here of symmetry breaking, but we must realize that in modern work the various causes which affect the equilibrium of the cone, such as the speed of the molecules in the air, the vibrations of the table, etc., can be considered symmetric only in an average statistical sense and that any fluctuation from this average will be a sufficient reason in the Leibniz sense for the equilibrium to break.

Figure 2 reveals a far more interesting, indeed intriguing case of (alleged) symmetry breaking. Imagine that the regular linear chain of atoms shown at

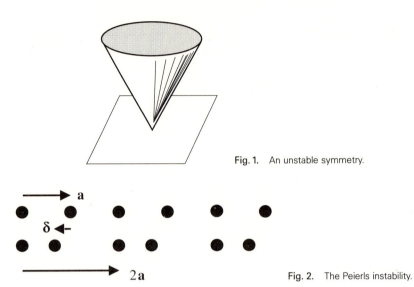

Fig. 1. An unstable symmetry.

Fig. 2. The Peierls instability.

the top of the figure extends to infinity at both ends. Then, if I make a copy of this infinite figure on a piece of transparent paper, I can translate the copy to the right or to the left by any integral multiple of the vector **a** shown and the copy and the original do not differ in any way at all. This is what I mean when I say that the translations by **a**, 2**a**, 3**a**, and so on are symmetry operations of the system. The rather extraordinary result was obtained by Peierls in 1955 that the regular linear chain shown will in many real cases spontaneously deform, forming alternating short and long bonds, as depicted at the bottom of Fig. 2, a phenomenon which is called the *Peierls instability.*

Notice that in the reconstructed linear chain the translations by **a** (and odd multiples) are no longer symmetry operations, so that symmetry has been lost in the process. We shall understand a little later on how such symmetry breaking is possible in apparent contradiction with our principle of symmetry conservation.

▋ A little history

So far in this book all the action has taken place outside mainland Britain, from Denmark to Ireland to France. We now go back to England, although the story starts in Berlin, where Rudolf Peierls was born in 1907. Peierls got his doctorate at Leipzig in 1929 and he then spent several years on post-doctoral fellowships that took him around Europe, from Leningrad to

Zürich, Rome, and Cambridge. By then, 1933 had arrived and with it the German troubles: Peierls' Jewish origin made a return to Germany unwise if not impossible, and he got a grant at Manchester University. From then on, although war work took him to Los Alamos, Peierls' academic life was centred in England. At 30 he was elected to the chair of Applied Mathematics (later renamed Mathematical Physics) at Birmingham, where he was to remain, occasionally delocalized however, for the next 25 years or so. In 1963 he became Wykeham Professor of Physics (a chair of theoretical physics) at Oxford, from which post he retired in 1973, having made the Department of Theoretical Physics there one of the most fruitful and enjoyable places at the university. In active retirement, he still lives at Oxford.

I am afraid that the discovery I shall tell you about is, for a physicist like Peierls, no more than peanuts. How else could it be for a man who was a co-signatory of the Frisch–Peierls Memorandum in 1940, which is the nearest thing to the first blueprint ever produced of the atomic bomb? Although it got him his knighthood in 1968, by that time Sir Rudolf had long realized the awesome implications of his invention and had already been for a number of years a leading force behind the Pugwash Conference movement.

Peierls' life is well documented in his autobiography (Peierls 1985) but, characteristically, there is nothing in it about the Peierls instability. The history of this discovery is somewhat unusual because the results were never published in a learned journal, as happens with most scientific work. I thought therefore that it would be interesting to record here a little more of the background of the discovery and I had a very pleasant discussion about it with Sir Rudolf. This is what he told me.

In 1953, when Professor Peierls (as he then was) was Professor of Mathematical Physics at Birmingham, he was invited to give a series of lectures on solid state physics at the Summer School at Les Houches, a condition of this invitation being that lecture notes had to be made available before the beginning of the School. In preparing those notes, and during the lectures, Peierls did not discuss the famous instability, but when he returned to Birmingham, his notes being fairly well developed, he decided to make them into a book, and it was during this process that he remembered some work done by Professor Harry Jones from Imperial College in London (about which more later on) and he then got the idea of the instability. Before going much further, he consulted with Maurice Pryce, the then Wykeham Professor at Oxford and, having received a favourable opinion, included the idea in his book, in a rather sketchy manner without going into details. So it is in this way that this important piece of work appeared in Peierls' book, published two years after the Les Houches School (Peierls 1955: pp. 108–12). This predicted effect was soon famous, since the Peierls instability plays an important role in the theory of the electrical conductivity of linear models, which became very important in the 1960s because of the advent of new polymers with possible technological applications.

The theory of the Peierls instability is in fact very simple, but it requires of me to review in something like a quarter of an hour all the major ideas of the band theory of solids, a good example of the impossible taking no time at all. It will be interesting, however, to try to do this, because it will give us a very good idea of how much one can get out of symmetry. I shall first have to discuss the way in which we are going to handle our linear chain model.

▮ The linear chain model

We must first of all understand the nature of the physical model of the linear chain which we must use in order to study its properties. We assume, to start with, that although each atom of the chain has a lot of electrons, most of these are tightly held near the atomic nuclei and contribute very little to the collective properties of the chain. Thus, we need be concerned with only a few electrons at each nuclear site, often no more than one or two, these being the more loosely bound electrons. These electrons, because they are much lighter than the nuclei and because they are loosely bound, move very rapidly, so that, *to a first approximation*, we can assume that the ion cores (nuclei plus the tightly bound electrons) are fixed and that the remaining electrons move in the field of these fixed cores. Just like electrons in atoms are always in one of a discrete set of energy states, these electrons will also be found in states with fixed energies. We are interested in these states because what happens to the linear chain, that is, the type of geometry which it will adopt, will depend on the total energy of this electron system: for different nuclear geometries this total energy will vary and whether we have a regular or an alternating chain will depend on which is the geometry that leads to the lowest total energy.

Although this first approximation will allow us (in principle) to compute the total energy of the electron system, it is too rough to be an adequate model of our linear chain. This is so because the nuclei vibrate, and vibrate more the higher the temperature: since we are interested in deformations of the chain in which the nuclei move away from their original positions of equilibrium, it would be ludicrous to ignore the vibrations of the system.

Rather than throwing away our first approximation of fixed nuclei, what our model entails is this. We first compute the electronic levels for fixed nuclei. Then, we consider the way in which the nuclei vibrate. Finally, we consider the interaction between the nuclear vibrations and the electronic states. As you can see, the idea is that we proceed in successive steps, each of which independently is rather over-simplified, but then, when putting everything together, a fairly good approximation is finally obtained.

Our total model must thus consist not just of a linear chain with its nuclei in fixed positions but also of a heat bath with which this chain interacts. The heat bath excites vibrational states of the chain, and these vibrational states interact with the electronic states, thus determining the final energy of the system (for each temperature of the heat bath, since at different temperatures different vibrational states will be excited). When we realize that this is the model that we have in mind and when we look at the icon of the linear chain at the top of Fig. 2, we realize how much we have to keep in the back of our minds when using the icon. Also, how easy it is to be mesmerized by the icon adopted for the linear chain and to forget all that lies behind it.

▌How to deal with translations

As I have announced, we now embark on a fifteen-minute course on the solid state, and the first thing we must learn is how to handle translations, which are the most characteristic symmetry operations in crystals. We shall consider for this purpose a linear regular chain of N atoms, where N is a very large number but, in order to have an easier example in front of our eyes, I show at the top of Fig. 3 a case where N equals 10.

We have already seen that in order to deal properly with translations we need an infinite chain. Things therefore become a bit complicated, for which reason one adopts a very drastic but nevertheless good approximation, which is to bend the chain so as to form a regular circle, as shown in the lower part of the figure. The justification for this so-called *cyclic condition* is that, if we have millions of atoms, the curvature introduced by forming the circle will be negligible and it will not change any physical

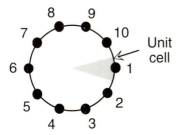

Fig. 3. The linear chain model and its cyclic form.

properties. One immediate advantage of this approach is that we have no trouble at all in recognizing here that we have now ten possible 'translation' operations, which are the rotations by $2\pi/10$ and its multiples from 1 to 10. Notice that under these rotations we can define a *unit cell*, as shown in the figure, such that 'translation' of this unit cell by all possible 'translations' reproduces the whole system.

Bear in mind that although we have in this case as many atoms as unit cells, this will not always be the case and that the important feature for us is that our system has N unit cells rather than N atoms. This is so because the unit cell is the entity that is repeated under the symmetry operation.

Now that we know all about the symmetry operations we have to think about their significance with respect to energies, and the question of the energy in these atomic systems is intimately connected with the concept of *degeneracy*.

Before we look into this, please do not feel cheated that I have promised to feed you with translations and that, in fact, what I have done is to put rotations on your plate. The physics of the model, as I said, allows you to do this and the mathematics is now ever so simple!

▌ Energies and degeneracies

We all know that in an atom we have a fixed nucleus and that there are electrons around it distributed amongst fixed energy states. One thought about this in the old days as if the electrons were like planets around the sun: the further the orbit, the higher their energy. We now know that such orbits do not exist but rather that the discrete energy states occupied by the electrons (now called *orbitals*) are characterized by precise probability distributions which give the probability of finding the electron at any time and at any point in space. The state designated by p_x in Fig. 4a is one such orbital, the required probability in any one direction being obtained from the length of the position vector of the corresponding point of the curve shown.

A very important property of the discrete energy levels that quantum mechanics finds in atoms and other systems is that they come in clusters of states, or wave functions, which all share the same energy: they are said to be *degenerate*. A very well known example is given by the so-called $2p_x$ and $2p_y$ states in atoms, shown in Fig. 4a. Why are they degenerate? The answer is very simple. An atom is spherically symmetrical, which means that any rotation about an axis passing through the nucleus is a symmetry operation and that it must therefore leave the energy invariant. (Symmetry operations, because they do not change the relations between any of the particles that

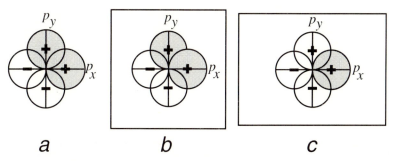

$$a \qquad b \qquad c$$

Fig. 4. Symmetries and degeneracies. All rotations described in the text are assumed to be about axes perpendicular to the drawing.

make up a system, cannot alter energies.) The two orbitals of the figure are degenerate because a 90° rotation will take one into the other and this operation is one of the infinite number of rotations that are symmetry operations of the atom.

If we place the atom in an environment that is not spherically symmetrical, things may change. First, consider the same orbitals in a cubic environment, such as that provided by a cubic crystal. Not all rotations are now permitted: the only ones allowed are those that leave the cube (or in Fig. 4*b* the square) invariant. The 90° rotation, however, is still there, so that the two orbitals in question are still degenerate. If we put the same atom in a non-cubic field, as in Fig. 4*c*, the 90° rotation is no longer a symmetry operation, and there is no symmetry operation of the system that will transform one orbital into the other: the degeneracy disappears and the single energy level we had splits into two, one corresponding to each state.

Having said all this about degeneracy many of you will ask: why on earth do we worry about this property? It is very easy to think that degeneracy is no more than a peculiar but perhaps unimportant feature of energy states. If one day, however, you were so fed up with the world that you wanted to destroy it, and a perverse genie granted you only two words to achieve your purpose, you could not say anything more efficacious than 'eliminate degeneracy', whereupon the whole universe would be vaporized out of existence. Why is this so?

Look again at the atom in Fig. 4*a*. The way in which electrons distribute themselves about the fixed energy states or orbitals is governed by the *Pauli principle*, which says that *each orbital* can accommodate no more than two electrons. You must please understand here that we do not say each energy state but rather each orbital: because p_x and p_y are degenerate they have the same energy, but we can now have four electrons at this energy rather than two. Since degeneracies can in practice be quite large, one can have, say, sixteen electrons rather than two at the same energy, and so on. If degeneracy did not exist the remaining fourteen electrons in this case would have to look for states of higher energy (because all the levels *below* the level which is being filled must be already occupied) and thus most atoms would have to have electrons with very high energies indeed. Imagine now what the situation would be in a solid with millions of electrons! So it is degeneracy that keeps us nicely as we are, and since degeneracy depends essentially on symmetry, you can see why symmetry is so important.

■ Classification of the energy states

Let us now go back to Fig. 4c and see how to classify the possible energy levels in accordance to the symmetry that we have. In order to make things easier I shall consider the single symmetry plane σ shown in Fig. 5, and we shall use the same Greek letter also to denote the operation of reflection through this plane. We can see that the two non-degenerate functions that we have behave differently under this operation:

$$\sigma p_x = +1 p_x, \qquad \sigma p_y = -1 p_y. \qquad (1)$$

We can actually *predict* that these are the two possible symmetry types generated by this operation from the following argument. If we repeat σ twice we do nothing in total. Thus, σ^2 must be the identity. Therefore,

$$\sigma^2 = 1 \quad \Rightarrow \quad \sigma \mapsto \pm 1 \quad \Rightarrow \quad \sigma\varphi = \pm 1\,\varphi. \qquad (2)$$

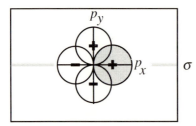

Fig. 5. Classification of symmetry types with respect to a reflection plane. The plane σ is perpendicular to the plane of the figure and to the y axis.

Notice that what we have done here is to show that, given the *single* symmetry operation σ, we must then have two non-degenerate states, one symmetric and the other antisymmetric with respect to σ, precisely as displayed in the figure and shown in equation (1). Just to make this as clear as we can: the last equation in (2) means that, when there is only one symmetry plane there are two possible symmetry types, one of which is called *symmetric*, corresponding to the + sign and for which the orbital could accordingly be labelled φ_+, and another called *antisymmetric*, corresponding to the − sign and for which the orbital could be labelled φ_-. These two orbitals cannot be degenerate because the only symmetry operation which we have, σ, is incapable of transforming one type into the other. (In accordance with equation (2) all that it can do is to leave each orbital as it is, except for multiplication by plus or minus one, depending on their symmetry type.)

Let us now try to apply the same principle to our linear chain in cyclic form, which we illustrate again in Fig. 6. What we have in mind is this. Whereas in an atom an electron moves around a fixed nucleus, we have in the linear chain a fixed framework of N nuclei and of, say, N or $2N$ electrons which move around this framework. By comparison with the atomic case, we expect these electrons to occupy discrete energy levels, each of which will correspond to some orbital φ. In the same manner as we have shown in (2) that a single symmetry plane leads to the existence of two possible orbital types, we must expect that the more involved symmetry of the linear chain will also allow us to predict which are the orbital types possible, and this is what we are now going to do. If we call N the number of unit cells, and C the rotation by $2\pi/N$, our basic 'translation', which takes one cell into the next, then, after N repetitions of this operation, we obtain the rotation by 2π, that is, the identity

$$C^N = 1. \tag{3}$$

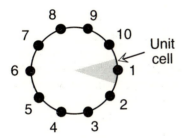

Fig. 6.   The linear chain in cyclic form.

In the same manner that the reflection σ multiplies an orbital φ by a factor ± 1, we want to find out the factor ω which corresponds to C:

$$C\varphi = \omega\varphi. \tag{4}$$

If we repeat N times the application of C on this equation, we get

$$C^N\varphi = \omega^N\varphi, \tag{5}$$

which, from (3), requires that

$$\omega^N = 1. \tag{6}$$

This means that ω must be one of the N roots of unity, ω_k:

$$\omega = \omega_k = \exp(-2\pi ik/N), \qquad k = 1, 2, \ldots, N. \tag{7}$$

(It is easy to verify, in fact, that ω^N equals unity.) The sign of the exponent is quite arbitrary here, but I keep it that way in order not to clash with various sensible conventions in the literature. What is important to remember is that, as displayed in (7), the number of such roots is precisely N, because if we add this number to k, no new root arises:

$$k' = k + N \quad \Rightarrow \quad \omega_{k'} = \omega_k \exp(-2\pi iN/N) = \omega_k. \tag{8}$$

We should rewrite our condition (4) a little more generally:

$$C\varphi_k = \omega_k \varphi_k, \qquad \omega_k = \exp(-2\pi ik/N), \qquad k = 1, 2, ..., N. \tag{9}$$

What we are saying here is this: just like a reflection plane allows us to classify the possible states as symmetrical and antisymmetrical with respect to it, the operation C allows us now to classify the possible states into N types φ_k, each type being characterized by one of the N roots of unity ω_k. (In order to be rigorous we would have to prove that the other symmetry operations, C^2, C^3, etc., do not introduce any new symmetry types, but this is a result which is not too hard to accept.)

The functions or orbitals that correspond to these symmetry types are called the *Bloch functions*, and they are the possible wave functions (that is, the functions that correspond to the stationary energy states) in our linear chain of N unit cells:

$$\text{Bloch functions:} \quad \varphi_k, \qquad k = 1, 2, ..., N. \tag{10}$$

Let us now look back at our linear chain model and refine it a little. If we consider Fig. 6 again, we must agree that each atom in it must be assumed at the start to be in the same atomic orbital φ^j, this being one of the atomic states, $1s$, $2s$, $2p_x$, etc., which correspond to the atoms of which the chain is made up. Naturally, they will be deformed when forming the solid, but it is nevertheless clear that, as in (10), each atomic orbital φ^j will produce a set of N Bloch functions which is called a *band*:

$$\text{Band:} \quad (\varphi^j)_k, \qquad k = 1, 2, ..., N. \tag{11}$$

It is useful to change the range of k in (10) and (11). Notice that k is a simple quantum number which denotes the energy state within a band, and that, because of the periodicity in k by N shown in (8), any N consecutive integers will do just as well as those given in those equations. Let us subtract $\frac{N}{2}$ from each of the values given there (I assume that N is even, which does not matter at all when N is large). The new range of k so defined is called the *Brillouin zone*:

$$\text{Brillouin zone:} \quad k = -\tfrac{N}{2}+1, \ -\tfrac{N}{2}+2, ..., \tfrac{N}{2} \qquad (N \text{ values}). \tag{12}$$

We have now done most of the fundamental work, but, before we go on, a warning and a summary will be useful. The warning is about N, which is not to be understood as the number of atoms (although this happens to be true in Fig. 6), but rather as the *number of unit cells*. This is important because a unit cell can have more than one atom and in all this work N was associated with counting all possible translations, a translation being always associated with unit cells rather than with the individual atoms in them. (Remember that at the bottom of Fig. 2 a translation by 2**a** corresponds to the transformation of the two atoms of the unit cell.)

Having issued this warning, we can now summarize our results. In order to make them easier to understand, we shall concentrate on the particular case when there is only one atom per unit cell. Thus, we have a linear chain with N atoms each of them with orbitals $1s$, $2s$, $2p_x$, $2p_y$, etc. Equation (11) tells us that from each of these N atomic orbitals precisely N orbitals or energy states for the linear chain can be formed. Each set of N orbitals from (11) will form successive bands (that is, bands with successively higher energies) which are the $1s$ band, the $2s$ band, and so on, the type of band depending on the original atomic orbital at each nuclear site. The N orbitals or energy states in a band are labelled by the *values of k in the Brillouin zone which are N in number* (see equations (11) and (12)).

Notice also that k labels a symmetry type with respect to translations. (This symmetry type is characterized by the N numbers ω_k in equation (9).) As follows from (8), both k and $k+N$ always label the same symmetry type with respect to translation, this being the reason why we have only N values of k in the Brillouin zone.

At each value of k, that is, for each symmetry type, there are lots of Bloch functions, corresponding to the first band, second band, and so on. This means that for the same value of k we have a succession of energy levels, for the successive bands. It might be possible, in principle, that two such levels might be degenerate, but this is not permitted for the following reason. For $(\varphi^1)_k$ and $(\varphi^2)_k$ to be degenerate we need a symmetry operation that turns one into the other, say

$$C(\varphi^1)_k = (\varphi^2)_k. \tag{13}$$

We know, however, from (9), that $C(\varphi^1)_k$ is $\omega_k(\varphi^1)_k$ which is merely a multiple of $(\varphi^1)_k$ and can never be the new function $(\varphi^2)_k$. (It is accepted in quantum mechanics that two functions that differ by a constant factor are for all uses and purposes the same function.)

In order to convince ourselves that this degeneracy between different bands at the same k cannot exist, we should consider all the possible symmetry operations in the system. It is easy to accept that no multiple of C can do anything useful in this respect. If, on the other hand, we look at our picture of the linear chain at the top of Fig. 3, and if we assume that it is extended to infinity, as we must, it is not difficult to see that there are centres of inversion. It is fairly easy, although I shall not try to do it, to verify that these centres of inversion are equivalent to the conjugator operator which we shall introduce a little later on (p. 82). In equation (13), however, we must keep k constant whereas we shall see that this conjugator operator always changes k into $-k$ (which it is not too hazardous to guess that the inversion will also do, this being the reason why it need not be considered).

Putting all these results together, we see that there is no symmetry operation that will change the Bloch function of one band into that of another band, *at the same k*, which is summarized in the following rule for the linear chain (see, however, p. 93):

> Two bands can never touch or cross.

In giving this rule, we imagine that the energy of each state in each band is plotted as a function of k in the Brillouin zone. Thus, if two bands were to cross or touch for one particular value of k, the two corresponding Bloch orbitals for each of the two different bands would have the same energy and thus be degenerate, which is not allowed, as we have seen.

It is this rule, in fact, that gives rise to the concept of *band*. Because of this rule, the energy levels in two successive bands have to be kept largely apart so that they actually tend to form bands of permitted energies along the energy axis. But we shall see this more clearly in the next section.

▎ Energy bands in the Brillouin zone

I show in Fig. 7 the results of a very simple calculation of the energy levels for the linear chain of ten unit cells (each with one atom) in Fig. 6. I have assumed one electron per atom and I have taken a convenient energy scale. The method I have used is a simplified form of the so-called *tight-binding method* in solid state, well known in chemistry as the *Hückel method*. What we expect to find from the work so far is that each band will contain ten energy states, and this is precisely what we find. From (12), the Brillouin zone will contain ten k values from -4 to 5, and we find indeed one state for each of these ten values. (Although, as is standard, we include -5 in order to symmetrize the Brillouin zone; we represent the corresponding energy level with a white circle since it should not be counted within the first band.) The energy scale on the right of Fig. 7 shows the energy levels in a way which is usual in chemistry, for instance when dealing with a molecule such as benzene.

Notice that we have joined the energy levels with a grey curve: if we were to increase the number of N from 10 to let us say 10 000, the edges of the Brillouin zone would be 5000 and -5000 (really, -4999!) and we would have 10 000 states in the first band, following very much the same curve. It is very easy these days to perform such an experiment on a desk-top

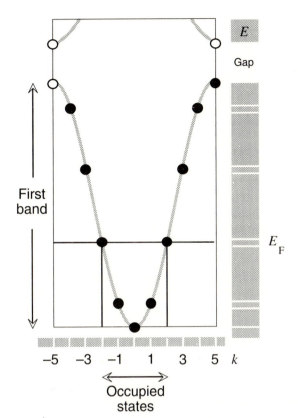

Fig. 7. The first band and part of the second band for the linear model of Fig. 6.

computer and it is quite amazing to see how accurately the curve is followed. The curve that one finds is very nearly a parabola, except near the edges of the Brillouin zone where it bends down as shown in the figure.

It must be remembered that in order to calculate a band such as the one displayed in Fig. 7 as the first band, one must assume that one given type of orbital is placed at each atomic site and the calculation which we have described is based on choosing for this purpose one s type orbital. If, instead, we place on each atom a p orbital like those displayed in Fig. 5, with the lobes along the direction of the chain, we would obtain another set of ten states, forming a second band, the lowest state of which is displayed at the top of Fig. 7. The curve that one finds this second band follows is very much like the first band, that is, it is very nearly a parabola over most of the Brillouin zone, except that it is inverted, going up rather than down from the Brillouin zone edge.

There are two rather remarkable results which these calculations show and which we must try to explain. First, except for k equal to 0 and 5, all levels (within the same band, of course) are doubly degenerate. Secondly, the two bands are well separated (as we have already suggested on p. 80 would be the case) with an *energy gap* at the edge of the Brillouin zone. We shall now try to explain these two results.

The degeneracy k, −k

We have seen in equation (9) that what makes a Bloch function be labelled as φ_k is the fact that it satisfies the relation

$$C\varphi_k = \exp\,(-2\pi ik/N)\,\varphi_k. \tag{14}$$

In order to explain the degeneracy that we have observed, we must find a symmetry operation which will transform φ_k into φ_{-k}. This will be amazingly easy and general, for there is an entirely new and very powerful symmetry operation which we can use, although it (at least at first sight) does not bear any relation at all to geometry. We start with the remark that, in a simple system as the one we are studying, every term which we may have in the energy will be real. (This, unfortunately, will no longer be true if spin is included explicitly, but we do not have to worry about such a case.) This means that we can take the complex conjugate of anything in our system and nothing significant will change: complex conjugation is thus a symmetry operation! In order to be able to handle this operation, let us define the conjugator operator K which acting on anything on its right transforms it into its complex conjugate. So, if w and z are complex numbers, for example $a+ib$ and $x+iy$ respectively, and $f(z)$ is a complex function of z (such as $\exp iz$, the complex character of this function coming from the presence of i *in the function*), then

$$Kwf(z) = w^* f^*(z^*). \tag{15}$$

(In the example given, the right-hand side here would be $(a-ib)$ times $\exp\{-i(x-iy)\}$, as can readily be seen.)

It is not difficult to prove that K commutes with all geometrical symmetry operations, a result which is basically a consequence of the fact that geometrical symmetry operations do not ever cause the conjugation of complex numbers or complex functions:

$$KC = CK. \tag{16}$$

Now apply K on both sides of (14):

$$KC\varphi_k = K \exp\,(-2\pi ik/N)\,\varphi_k. \tag{17}$$

Commute the operators here on the left, as in (16), and apply (15) on the right:

$$C(K\varphi_k) = \exp\,(2\pi ik/N)\,(K\varphi_k). \tag{18}$$

The brackets here around $K\varphi_k$ are purely cosmetic, but they help us to see that this function satisfies the defining equation of the Bloch functions, equation (9), for the value $-k$ instead of k. Since this label of the Bloch function depends entirely on the value of ω in (9), $K\varphi_k$ should be labelled as φ_{-k}:

$$K\varphi_k = \varphi_{-k}. \tag{19}$$

Thus, φ_k is transformed into φ_{-k} by the symmetry operation K, which explains why these two functions are degenerate, as we have seen in Fig. 7, where the double lines on the energy scale emphasize this degeneracy. Naturally, when k and $-k$ differ by N, as happens at the edge of the Brillouin zone, they must be considered identical (see equation (8)), so that the degeneracy does not exist because the two functions φ_k and φ_{-k} coincide. (This is why, as implied by Fig. 7, φ_k, for k equal to $-\frac{N}{2}$, is not included in the band.)

The band gap

The second feature of Fig. 7 which we must explain is the existence of the band gap at the edge of the Brillouin zone, k equal to $\frac{N}{2}$, and we shall do this as well as we can, but the reader cannot expect more than a sketch of an argument since a full treatment of this problem is rather involved.

We show in Fig. 8a the top of Fig. 7, where in fact a band gap appears between the first and the second bands. Remember that the atomic orbital assumed in the first band is $1s$, whereas in the second band we start from a $2p$ orbital at each site, which is the reason why the levels of the second band are of higher energy than those of the first band. The origin of the gap is now seen to be a direct consequence of the rule boxed on p. 80. If we were to assume, for instance, that the second band is shifted down as shown in Fig. 8b, then the two bands cross, which is not permitted. Neither would it be allowed for the second band to be moved down so as just to touch the first band at the Brillouin zone edge. From this point of view the energy gap at the edge of the Brillouin zone is a vestigial trait of the quantum jumps that separate the atomic levels.

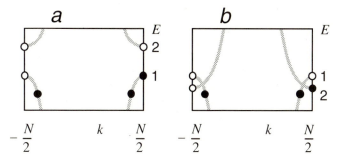

Fig. 8. The band gap.

It must be strongly emphasized that this argument is valid only for a linear chain as treated here. This is so because the non-crossing rule of p. 80 depends on the fact that *in this case* there is no symmetry operation which could introduce a degeneracy between two Bloch functions at precisely the same k. We shall see, in fact, that when more sophisticated symmetries are considered the situation does indeed change.

The Fermi energy

We now understand all the details of Fig. 7 and we can discuss the electronic structure of our linear chain. Let us first remind ourselves of how one works out the eletronic structure of, say, the sodium atom. It is known that there are 11 electrons in this atom which must be distributed among the permitted atomic states, which are, in order of ascending energy, $1s$, $2s$, $2p_x$, $2p_y$, $2p_z$, $3s$, $3p_x$, and so on. The Pauli principle does not allow us to place more than two electrons (with opposite spins) in each state, and we must start filling states from the bottom up. We thus place two electrons each in the first five states, $1s$, $2s$, $2p_x$, $2p_y$, $2p_z$, and we are left with one electron to place in the $3s$ level, which is thus the top occupied level. Knowing, as in this case, which is the top occupied level will tell us a good deal about the electronic energy of the atom.

Let us now consider in the same way the electronic structure of the linear chain of Fig. 7. Because we have ten atoms each with one electron, we have to accommodate ten electrons in the energy levels shown in the figure. In doing this we must make a distinction between levels and states: in Fig. 7, for instance, the second level, because of its degeneracy, corresponds to two states. This is important because again the famous Pauli principle requires that no more than two electrons be placed in the same *state*. Thus, because the first level in Fig. 7 contains only one state, it can take only two electrons, but because the second and third levels each corresponds to two states, they can take four electrons each. We can see at once that we can place all our ten electrons by filling all states up to and including the third level shown in the figure.

The top occupied level gives a fairly good indication of the total electronic energy of the system and it is called the *Fermi level* or the *Fermi energy*. Because there are plenty of energy levels slightly above the Fermi energy in Fig. 7 (we are assuming that N is large and thus that we can follow the grey curve) the system of electrons can be excited by an electric field and it is this which causes electrical conduction. (It must be realized that the normal range of electric fields available in practice is such that the energy that they can give to an electron is very small indeed in terms, say, of our vertical scale in Fig. 7.) If, instead, the whole Brillouin zone were full, this excitation could not happen, because there are no energy states in the band gap and because the kick given to the electrons by the field is not strong enough to push them up to the second band. In this case, conduction would not arise because the electrons are unable to receive energy from the electrical field, which they can only do if they can jump up to a higher energy state.

We now know everything we need about solid state theory and we can understand the Peierls effect. Before we do this it is useful to summarize and simplify some of the properties of the Brillouin zone which we have used. Notice that, if N is large, we can ignore the difference that unity makes in equation (12) and thus happily say that the Brillouin zone extends from k equal to 0 up to $\pm \frac{N}{2}$. It should be clear that $2N$ electrons will occupy all the N states in the Brillouin zone of (12) or of our now slightly doctored Brillouin zone. On the other hand, N electrons will occupy $\frac{N}{2}$ states in this zone which, again for large N, can be taken to range from k equal to 0 up to $\pm \frac{N}{2}$. (The reader can easily verify *approximately* these statements in Fig. 7 where N, of course, is not sufficiently large.) Notice also from this figure that the length of any chosen region of the Brillouin zone determines the number of electron states in it. All these properties will be very important in our later discussion.

The Peierls instability for the linear chain model

Our problem is the following one: given the regular linear model at the top of Fig. 9 (assumed extended to infinity) we want to know whether the corresponding alternating structure shown at the bottom of the figure will have lower energy, in which case the chain will deform in order to adopt this more stable structure. We assume, as always, that we have only one electron per atom, that is, that altogether we have N electrons.

The answer to this question is easily provided by Fig. 10. The important feature of the transition to the alternating structure which is revealed by Fig. 9 is that the total number of unit cells is halved from N to $\frac{N}{2}$. In Fig. 10a we have exactly the same situation as in Fig. 7, since we are plotting the first

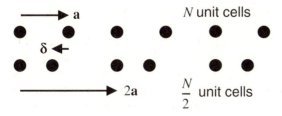

Fig. 9.   The regular linear chain model and its alternating form.

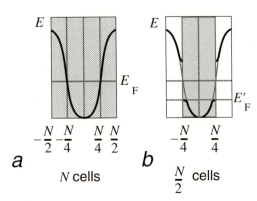

a

N cells

b

$\frac{N}{2}$ cells

Fig. 10.   The Peierls instability. The shaded areas indicate the extent of the Brillouin zone along the horizontal axis.

band for the regular linear chain with N unit cells. Accordingly, as we have just seen, the Brillouin zone goes from $-\frac{N}{2}$ to $\frac{N}{2}$, and it is occupied by N electrons from $-\frac{N}{4}$ to $\frac{N}{4}$. In the alternating structure, on the other hand, the number of unit cells is $\frac{N}{2}$, so that the Brillouin zone must go from $-\frac{N}{4}$ to $\frac{N}{4}$. We have seen, on the other hand, that there is a band gap at the edge of the Brillouin zone, so that this must now appear in the alternating chain at $\frac{N}{4}$, as shown in the picture.

It is clear that the total energy in either case depends on the Fermi energy, which is very easy to obtain. We have precisely the same number of electrons in either case, N, so that precisely the same region of the horizontal axis must be occupied, from $-\frac{N}{4}$ to $\frac{N}{4}$ (that is, altogether $\frac{N}{2}$ states with two electrons each). Because of the gap at the newly formed Brillouin zone edge in Fig. 10b, it is clear that the new Fermi energy must be lower than in the regular structure and it is this effect which makes the alternating structure more stable.

We must now try to understand how it is that the symmetry of the regular structure is broken by the Peierls transition, since we appear to violate the principle of symmetry conservation. It seemingly follows from Fig. 9, in fact, that if we start with the chain in the regular structure shown in there, then that regular structure must be kept, since, otherwise, symmetry would be lost. It would be foolish, though, to be surprised if the Peierls transition does take place, since the argument so far has been concerned with the purely electronic structure which must be associated with the chains, and as we have seen when discussing the physical model of the linear chain (see p. 72) there are other terms in our physical model that must be taken into consideration.

It is a fact, however, that people do get surprised by the Peierls result, and they in a way express this surprise by saying that the symmetry of the regular linear chain is broken. We have to be careful here: if, when the proverbial stage magician pulls a rabbit out of his hat, the audience is surprised, it is good; but if it is the magician who is surprised, then the magician is a fool! When we say that symmetry has been broken we must realize that our metaphorical hat, which for us is our regular icon, was not at all empty. This icon depicts only *part* of our model: as we have explained already, we must consider the linear chain, as depicted by our icon, in contact with a heat bath. It is easy to think that if we subsume this icon in a heat bath, because the latter can be assumed to be totally symmetrical, then the symmetry of the icon is still valid. We must, however, consider a third stage in our model by including the interactions between the chain as represented by the icon and the heat bath, and it is these interactions that need not be symmetrical, as we shall now see.

Because the atoms of any real linear chain must be in contact with some heat bath, however cool, it means that the chain must vibrate. When the vibrations of a solid are studied one finds that, in the same way as the electronic energy levels themselves, the vibrational levels are discrete and that a given system will admit of a finite number of so-called *normal modes of vibration* such that they and their harmonics form all the possible vibrational states of the system. In each of these normal modes the atoms move one with respect to the other in very well determined relations, and one such normal mode is shown at the top of Fig. 11 for the regular structure of the linear chain. It is clear that when the regular structure starts to vibrate in this particular mode it will go through an alternating form, as shown at the bottom of the figure. (Notice that this will not be the case for other vibrational modes.) Since, as follows from the Peierls argument, this alternating form is lower in energy, the system now becomes trapped in this configuration.

When we say that symmetry is broken in the Peierls transition, we are simply being mesmerized by the strong symmetry of the regular chain icon. This icon, however, is only a very rough representation of the physical model that we must adopt for our system. The model of the system that we must have in mind can be considered to be the linear chain plus the ensemble of all its possible vibrations, the latter replacing in our model the heat bath. When we do this we no longer have the symmetry we assumed at the beginning, that is, the symmetry of our crude icon, because the inter-action between the electronic states of the chain and the ensemble of its normal modes of vibration is not symmetrical: the interaction with the vibration shown in Fig. 11, for instance, is favoured over all others because it leads to a lower total energy, as we have shown. The problem therefore is that we start with a crude icon, the symmetry of which is not really physically significant, and when the description of the physical system is

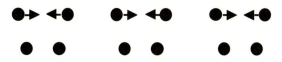

Fig. 11. The vibration of the linear chain (normal mode) which drives the Peierls transition.

improved in a further approximation, then the *non-physical symmetry* of the icon is 'broken'.

■ The quasi-linear chain

Although the linear chain of Fig. 9 is important since it has given us an idea about the effect of the Brillouin zone edges in the question of phase stabilities (a mechanism which had already been discussed before the war by Professor Harry Jones, and which Peierls had in mind when discussing his effect), genuine linear chains are of limited practical importance, since they are not often found in nature. If we consider a polyacetylene polymer, a material of considerable technological interest, we can see that it forms a zigzag chain (Fig. 12). In Fig. 12*a* we show the regular structure with all the carbon–carbon bonds of the same length in which, in order to keep to the four-valent character of the carbon atom, we assume that there is a so-called π orbital at each atom and normal to the plane of the figure. It is this fairly mobile electron in which we shall be interested, since we expect that it will form a mobile, conducting band. The possibility of alternation in this case must be considerable, since another structure of the polymer is possible in which we have alternating double (short) and single (long) bonds, as shown in Fig. 12*b*. It is thus very important to study the possibility of a Peierls transition in this case.

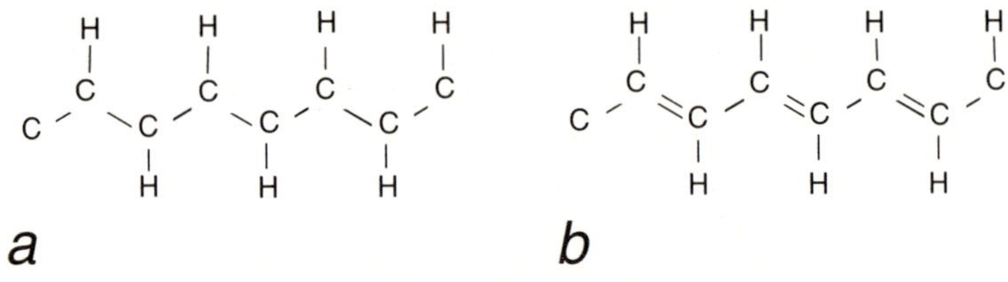

Fig. 12.   Parts of a polyacetylene polymer chain.

We depict the corresponding models for this structure in Fig. 13 and we immediately see that the Peierls transition to an alternating structure does not entail in this case any change at all in the number of unit cells, whereas, before, this number was halved. Even in the regular structure there are two atoms per unit cell and this number does not change on alternation. What happens, instead, is that one symmetry operation of a new type which exists in the regular chain, and which is called a *glide plane G*, disappears in the alternating structure.

Let us explain this. Notice in Fig. 13*a* that σ is not a symmetry plane, nor is the translation by $\frac{1}{2}\mathbf{a}$, denoted $C_{1/2}$, a symmetry translation. When these

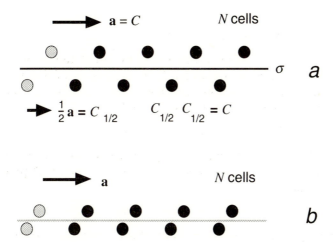

Fig. 13.   The quasi-linear chain model. The unit cells are indicated by the atoms in grey.

two operations are combined, however, a new symmetry operation is obtained, the glide plane G:

$$G = C_{1/2}\sigma. \tag{20}$$

This can easily be verified by reflecting a copy of the system on tracing paper and then translating this copy by $\frac{1}{2}\mathbf{a}$, whereupon covering of the system will be obtained. (Remember that these models must always be assumed to extend to infinity on both ends.) It is pretty clear on the other hand that G no longer is a symmetry operation of the alternating structure.

The whole mechanism of the Peierls instability in the quasi-linear chain depends on the entirely new fact that there is a glide in the regular quasi-linear chain which, as we shall see, introduces a special degeneracy whereby the two levels at the Brillouin zone edge corresponding to the first and second band (which are normally separated by the energy gap) become degenerate so that the gap disappears. In the alternating structure, on the other hand, there is no glide, the new degeneracy is split and the gap is brought back. We shall find that this result immediately requires the alternating structure to be the more stable of the two.

I shall now prove these important results and many readers, I am afraid, will find the arguments that follow somewhat prolix. The problem is that it is very easy to talk rubbish when discussing degeneracy and, even if one is not aiming at much rigour, a minimum of decency in the treatment is absolutely essential if one is not to fall into nonsense. It should be possible, nevertheless, for the reader to follow the gist of the argument without getting too involved in the details of the proofs provided. Before we start on these proofs it will be necessary to consolidate our ideas of degeneracy, as well as to collect together a few useful properties of Bloch functions.

Bloch functions and degeneracies

I shall first revise some of the properties of the Bloch functions in order to refine a little our hitherto somewhat loose discussion of degeneracy. The Bloch function φ_k is a function such that when acted upon by the translation C, it is multiplied by the factor $\exp(-2\pi ik/N)$:

$$C\varphi_k = \exp(-2\pi ik/N)\,\varphi_k. \qquad (21)$$

In discussing Bloch functions we might imagine that, whenever there is another function, φ_k' say, that satisfies the same equation with precisely the same coefficient,

$$C\varphi_k' = \exp(-2\pi ik/N)\,\varphi_k', \qquad (22)$$

then φ_k' and φ_k are necessarily *distinct* Bloch functions corresponding to the same k as, for instance, Bloch functions of different bands at the same k are. This is not true. If we consider equation (21) we must realize that if α is any constant, then φ_k multiplied by α still satisfies the same equation:

$$C\varphi_k = \exp(-2\pi ik/N)\,\varphi_k \quad \Rightarrow \quad \alpha C\varphi_k = \alpha \exp(-2\pi ik/N)\,\varphi_k$$

$$\Rightarrow \quad C\alpha\varphi_k = \exp(-2\pi ik/N)\,\alpha\varphi_k. \quad (23)$$

(We exploit here the fact that α is a constant and thus that $C\alpha\varphi_k$ is the same as $\alpha C\varphi_k$.) This means that if we have another function φ_k' which satisfies the same relation (21) as φ_k, then this function can in principle be a multiple of φ_k,

$$C\varphi_k' = \exp(-2\pi ik/N)\,\varphi_k', \qquad \varphi_k' = \alpha\varphi_k. \qquad (24)$$

We do not say in this case that φ_k and φ_k' are distinct Bloch functions because the relation between them is trivial: they are practically the same thing. (It is, in fact, a well known property of quantum mechanics that a function that describes a state can be multiplied by any constant without change in its physical meaning.)

The simple argument just made will teach us an important idea about degeneracy. Since φ_k and φ_k' are, as we have just said, the same function for all uses and purposes, they must both correspond to the same energy and thus we might be inclined to think that they are degenerate. *This is not so.* In order to have degeneracy we must insist that the two functions be quite independent of one another, as, for instance, the famous p_x and p_y orbitals are: although they are somehow related, of course, one is never simply a multiple of the other.

The next fact that we must remember is that, from equation (12), the Bloch functions within a band are labelled by the N values of k from $-\frac{N}{2}+1$ (effectively $-\frac{N}{2}$) to $\frac{N}{2}$. We also know that these functions come in degenerate pairs, the members of each pair being linked by the conjugator operator

$$K\varphi_k = \varphi_{-k}. \tag{25}$$

This degeneracy, however, fails to exist for two points of the Brillouin zone. One will be the origin, k equals zero, for which no negative partner exists. The other will be the edge k equal to $\frac{N}{2}$. This is so because for this value of k its partner φ_{-k} coincides with it, as we shall now show. We know from equation (8) that N can always be added to any k without affecting its meaning. Thus $-k$ is the same as $-k$ plus N:

$$k = \tfrac{N}{2} \quad \Rightarrow \quad -k = -\tfrac{N}{2} = -\tfrac{N}{2} + N = \tfrac{N}{2} = k \quad \Rightarrow \quad K\varphi_k = \varphi_{-k} = \varphi_k. \tag{26}$$

Thus, for this particular value of k at the edge of the Brillouin zone the degeneracy introduced by the conjugator disappears.

The Bloch function corresponding to the Brillouin zone edge has a second important property which we must bear in mind: it changes sign when acted upon by the translation C, as results from equation (21):

$$k = \tfrac{N}{2} \quad \Rightarrow \quad C\varphi_k = \exp\left(-2\pi i \tfrac{N}{2}/N\right)\varphi_k = \exp\left(-\pi i\right)\varphi_k = -\varphi_k. \tag{27}$$

The degeneracy introduced by the glide plane

We now come to the meat of our argument. For the true linear chain we saw in relation to equation (13) that there is no symmetry operation that converts a Bloch function for that k into another distinct Bloch function for the same k. This property is important because it means that you cannot have two degenerate Bloch functions at the same k. In the quasi-linear chain, on the other hand, this result will be breached for the Bloch function φ_k at the edge of the Brillouin zone:

$$\varphi_k, \qquad k = \tfrac{N}{2}, \tag{28}$$

so that an entirely new degeneracy, with momentous consequences, will appear at this edge. In order to prove that this is the case, we must find a symmetry operation which must have two wonderful properties. One is that acting on φ_k from (28) it does not change its k value. The other is that the function so generated is truly distinct from the original function. We shall assert that this symmetry operation is given by the product of the glide G from (20) with the conjugator K. We thus realize at once why this situation must be specific to the quasi-linear chain since the crucial factor is that we now have an entirely new operation, the glide.

Since G is a symmetry operation and so is K, then $GK\varphi_k$ can in principle be degenerate with φ_k. The first property we have to prove is that it belongs to the same value of k as φ_k does. (Remember that everything we now do is valid only for k at the edge of the Brillouin zone.) This means that it can be written as follows:

$$GK\varphi_k = \varphi_k'. \tag{29}$$

I shall not prove this result, since basically it is nothing more than (26). Equation (29) does not guarantee that a degeneracy exists between φ_k and φ'_k, because they could be trivially related, as in (24). We must therefore prove our second wonderful property, namely that φ_k and φ'_k are distinct, which means for us, as discussed in relation to equation (24), that one must not be a multiple of the other:

$$\varphi'_k \neq a\varphi_k. \tag{30}$$

We shall do this by reduction to the absurd. The proof that follows is quite elementary but readers will not lose the thread of the argument if they skip it. Suppose that the equality sign holds here. Then, from (29) and (30),

$$GK\varphi_k = a\varphi_k. \tag{31}$$

We must prove that such an a cannot exist, for which purpose we act on both sides of this equation with GK:

$$GK\,GK\varphi_k = GKa\varphi_k = a^*GK\varphi_k = a^*a\varphi_k. \tag{32}$$

(We use here the definition of the conjugator as given by equation (15).) On the left-hand side of this equation we can apply the already stated property that the conjugator commutes with all geometrical symmetry operations:

$$GK\,GK\varphi_k = GGKK\varphi_k. \tag{33}$$

There are two things we can do here. First, from the definition (20) of the glide, we have

$$GG = C_{1/2}\sigma C_{1/2}\sigma = C_{1/2}C_{1/2}\sigma\sigma = C, \tag{34}$$

on using the result for the square of $C_{1/2}$ shown in Fig. 13, and the property of the reflection that it gives the identity when repeated twice. The second result we can use in (33) is that the square of K must be the identity, since two successive conjugations cancel each other. Equation (33) therefore takes the form

$$GK\,GK\varphi_k = C\varphi_k = -\varphi_k, \tag{35}$$

from (27). On comparing (32) with (35), we get

$$a^*a = -1, \qquad \text{impossible}, \tag{36}$$

since the left-hand side here is the sum of two squares which cannot be negative. (Remember that if a is $m+in$, then a^*a is $(m-in)$ times $(m+in)$, which equals m^2+n^2.) Thus (30) follows, which means that φ_k and φ'_k are degenerate.

Brillouin zone and bands for the quasi-linear chain

We must now think a little about the meaning of the new degeneracy just found. We must remember that we are dealing here, not with an arbitrary value of k, but rather with the value of k at the Brillouin zone edge. For a general k value we must have the same situation as in the linear chain. This means that there cannot be two Bloch functions degenerate at that precise value of k, that is, that two bands cannot touch or cross there. At the edge of the Brillouin zone, in contrast, our new result means that we have two Bloch functions belonging precisely to that value of k which are nevertheless degenerate. If we look at Fig. 7 the only two Bloch functions we can have corresponding to this value of k are those of the first and second band respectively. These are therefore the functions that are degenerate, which means that the energy gap between the first and second bands must close, as shown in Fig. 14. Notice that we have drawn the second band so that it is symmetric in k, in order to give the correct degeneracy, including the fact that the state with k equal to zero is non-degenerate.

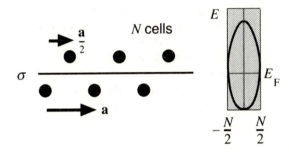

Fig. 14. The regular quasi-linear chain and its energy bands. The shaded area indicates the extent of the Brillouin zone.

Let us look at the way in which we have filled the states here to get the Fermi energy. Since each cell has two atoms and each atom has one electron, we have to place $2N$ electrons, for which we have to fill *all* the states in the Brillouin zone thus occupying the whole of the first band (see p. 85). We have obtained in this way the Fermi energy shown in the figure. Notice that, had it not been for the degeneracy just discovered, we would have had an energy gap at the Fermi level, which would make the material an insulator. It is, in fact, because of the result shown in this figure that the regular polyacetylenes can conduct.

When the chain is reconstructed to form short and long bonds, the glide plane is lost, and with it the degeneracy at the edge of the Brillouin zone disappears, so that the bands change, creating the usual gap at the Brillouin zone edge, as shown in Fig. 15.

Two results are immediately clear, since the way in which the Brillouin zone is occupied is exactly the same as before (same number of unit cells and of electrons). First, the Fermi energy goes down, so that the structure shown in this picture must be the favoured one. Secondly, there is now an energy gap at the Fermi level which means that the material ceases to conduct. It will be appreciated how important the Peierls instability is if we

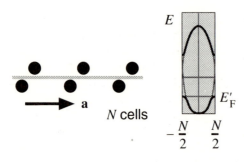

Fig. 15. The alternating quasi-linear chain and its energy bands. The grey lines and curves in the bands picture correspond to the regular chain.

want to have linear conductors. The situation appears hopeless, since we seem always to move towards the alternating, non-conducting state. It must be appreciated, nevertheless, that this is the favoured structure only from the point of view of the electronic energy. We must, however, spend some vibrational energy for the change to take place and, although it can be proved that, in general, this is second order with respect to the electronic energy, it is possible that the balance between the total electronic energy and the vibrational energy required for the deformation be such that the alternation predicted by the Peierls effect does not take place, a situation which will in general depend on the temperature of the material. It is thus possible that both structures are stable at different temperature ranges, separated by a temperature at which a Peierls transition takes place.

■ Conclusion

I hope that we have learnt a few things from this chapter. The first important idea is that symmetry is intimately connected with degeneracy. Degeneracy provides the guiding rules for the filling of energy levels in atoms and solids so that the total electronic energies of these systems depend crucially on the degeneracies of the electronic states in them. It follows that the study of symmetry provides a remarkable insight into the nature of the energy properties of quite complex systems.

We have also learnt how to classify all the energy states in a periodic system by means of a wonderful graphical device, called the *Brillouin zone*, each point of which corresponds to a symmetry type with respect to the translation operations of the system. Because of the intimate relation between symmetry and degeneracy, the Brillouin zone allows us to do most simply two very important jobs. One is to count the total number of states that we have and the other is to classify very rapidly, graphically, their degeneracies. It can thus be determined in what way the various electronic states possible are actually occupied by electrons and which is the highest occupied level of Fermi energy of the system. Knowledge of this Fermi energy thus gives a very good idea of the magnitude of the total electronic energy of the system, which can be used to decide which of two competing geometrical structures of a system would be energetically more favourable. (At a better level of approximation, however, one would have to study the mechanical energy necessary for the possible distortion of the system, as well as a variety of other higher order terms.)

One of the major themes of this chapter was the problem of symmetry breaking as opposed to that of symmetry conservation. In a way, we met with a symmetry paradox since it turned out that a perfectly regular linear chain would reconstruct into an alternating form in which symmetry appeared to be spontaneously lost. We recognized, however, that to speak of symmetry breaking here results from a misconception. The symmetry that is broken is that of the icon of the linear chain, but this icon depicts only one aspect of the total physical model which must be used in order to treat the system realistically. The total physical model, in fact, does not possess the full symmetry manifested by the icon used. Remember also, as it should be clear from Chapter 1, that the symmetry of an icon is not necessarily physically significant, so that breaking its symmetry, as opposed to breaking the symmetry of a physical system, is nothing worth writing home about. And it goes without saying that one must always avoid the trap revealed in Chapter 2, whereby one might read the icon of one object as if it was the icon of an entirely different object, in which case symmetries would get hopelessly confused.

The main thrust of Chapter 1, which is the distinction between a physical object and its icon cannot be sufficiently stressed. It is only too easy to draw an arrow and to say 'this is a vector' and then to proceed to draw physical conclusions from the sign used instead of going back to the object signified. This is as wrong as trying to place a cup of tea on the word 'table', thus confusing a sign with its signified. Whereas this type of behaviour is rare outside lunatic asylums, the traditions of science have not always kept its practitioners from iconic aberration, as our stories have illustrated.

If all these pitfalls are avoided, the good old principle of symmetry conservation is still pretty healthy. We must appreciate nevertheless that quantum mechanics has dealt a blow to Leibniz's principle of sufficient reason, since random events may take place for which no sufficient reason can be found and which can indeed produce genuine symmetry breakings. Before such ultimate events must be invoked, it is important, however, to examine carefully the suppositions involved by the icons and models used: it is quite possible that the explanation of why a rabbit came out of a hat is simply that a rabbit had been put there in the first place. Look for it!

∎ Epilogue

I should like now to tidy up a few historical points. The first one is somewhat disturbing: it is often the case, however, that good ideas are invented not once but at least twice and the Peierls instability is in fact a particular case of a very famous effect called the Jahn–Teller effect. (It must be coincidental, I hope, that Peierls and Teller invented respectively the atom and the hydrogen bombs!) Although Jahn and Teller (1937) preceded Peierls by a good eighteen years, they did their work in relation to molecules and thus, although the effect named after them was well known to chemists, it was not the sort of thing that physicists were very much interested in by the time Peierls came into the picture. Teller himself (see Teller 1972) later gave credit for the idea of the Jahn–Teller effect to Landau, who had discussed it with him in Copenhagen in 1934 but who did not bother to go through the tedious group-theoretical verification, symmetry group by symmetry group, which was subsequently done by Jahn and Teller. Modestly, Teller writes: 'Jahn and I merely did a bit of spade work.'

You will remember that before publishing his book in 1955 Professor Peierls consulted with Professor Pryce, who then went on to publish a seminal paper on the Jahn–Teller effect in 1957 (see Öpik and Pryce 1957). One wonders whether there was something about the Peierls effect that led Professor Pryce to look into the Jahn and Teller work. The fact is that the relation between these two important effects, however, was somewhat neglected for quite some time. Salem (1966) was one of the first to redress the balance.

The remarkable relation between glide planes and degeneracy was first studied in a much forgotten paper by Hund (1936). The polyacetylenes, in which, as we have seen, such glides are crucial, have been extensively studied with somewhat contradictory results. Longuet-Higgins and Salem (1959), quite independently of Peierls, concluded, as suggested by the Peierls effect, that the alternating structure was the most stable one. When the accuracy of the model was improved by Harris and Falicov (1969), the possibility of alternation was nevertheless shown to be negligible. More references to the literature, as well as details of the missing proofs in this chapter, may be found in Chapter 12 of Altmann (1991).

References

Alexander, H. G. (1956). *The Leibniz–Clarke Correspondence.* Manchester University Press.

Altmann, S. L. (1986). *Rotations, Quaternions, and Double Groups.* Clarendon Press, Oxford.

Altmann, S. L. (1989). Hamilton, Rodrigues, and the Quaternion Scandal. *Mathematics Magazine,* **62**, 291–308.

Altmann, S. L. (1991). *Band Theory of Solids: An Introduction from the Point of View of Symmetry.* Clarendon Press, Oxford.

Aquinas, St Thomas (1856). *Summa Theologica* (15th edn). Bloud et Barral, Paris.

Cartan, E. (1913). Les groupes projectifs qui ne laissent invariante aucune multiplicité plane. *Bulletin de la Société Mathématique de France,* **41**, 53–96.

Chalmers, T. W. (1949). *Historic Researches. Chapter in the History of Physical and Chemical Discovery.* Morgan Brothers, London.

Curie, P. (1894). Sur la symétrie dans les phénomenes physiques, symétrie d'un champ électrique et d'un champ magnétique. *Journal de Physique Théorique et Appliquée,* 3e série, **3**, 393–415.

Dibner, B. (1961). *Oersted and the Discovery of Electromagnetism.* Burndy Library, Norwack, Connecticut.

Ellmann, R. (1987). *Oscar Wilde.* Hamish Hamilton, London.

Feynman, R. P., Leighton, R. B., and Sands, M. (1963). *The Feynman Lectures on Physics.* Addison-Wesley, Reading, Massachusetts.

Giedymin, J. (1982). *Science and Convention. Essays on Henri Poincaré's Philosophy of Science and the Conventionalist Tradition.* Pergamon Press, Oxford.

Goodman, N. (1965). *Fact, Fiction, and Forecast* (2nd edn). Bobbs-Merrill, Indianapolis.

Graves, R. P. (1882). *Life of Sir William Rowan Hamilton* (Vol. 2, 1885; Vol. 3, 1889). Hodges, Figgis & Co., Dublin.

Hamilton, W. R. (1853). *Lectures on Quaternions.* Hodges & Smith, Dublin.

Hankins, T. L. (1980). *Sir William Rowan Hamilton.* Johns Hopkins University Press, Baltimore.

Harris, R. A. and Falicov, L. M. (1969). Self-consistent theory of bond alternation in polyenes: normal state, charge-density waves, and spin-density waves. *Journal of Chemical Physics,* **51**, 5034–51.

Heath, T. L. (1897). *The Works of Archimedes.* Cambridge University Press.

Hund, F. (1936). Über den Zusammenhang zwischen der Symmetrie eines Kristallgitters und den Zuständen seiner Elektronen. *Zeitschrift für Physik,* **99**, 119–36.

Institut de France, Académie des Sciences (1916). *Procès-Verbaux des Séances de l'Academie tenues depuis la fondation de l'Institut jusqu'au mois d'août 1835.* Tome VII, Années 1820–1823. Imprimerie de l'Observatoire d'Abbadia, Hendaye.

Jahn, H. A. and Teller, E. (1937). Stability of polyatomic molecules in degenerate electron states. I – Orbital degeneracy. *Proceedings of the Royal Society,* **A161**, 220–35.

Jones, B. (1870). *The Life and Letters of Faraday* (2nd edn). Longmans, Green & Co., London.

Kirk, G. S., Raven, J. E., and Schofield, M. (1983). *The Presocratic Philosophers* (2nd edn). Cambridge University Press.

Kopstik, V. A. (1983). Symmetry principles in physics. *Journal of Physics C: Solid State Physics*, **16**, 23–34.

Leibniz, G. W. (1951). *Theodicy* (transl. E. M. Huggard). Routledge & Kegan Paul, London.

Longuet-Higgins, H. C. and Salem, L. (1959). The alternation of bond lengths in long conjugated chain molecules. *Proceedings of the Royal Society*, **A251**, 172–85.

Mach, E. (1893). *The Science of Mechanics: A Critical and Historical Exposition of its Principles* (trans. T. J. McCormack, from the 2nd German edn; 1st German edn, 1883). Watts & Co., London.

Nagel, E. (1961). *The Structure of Science. Problems in the Logic of Scientific Explanation.* Routledge & Kegan Paul, London.

O'Donnell, S. (1983). *William Rowan Hamilton. Portrait of a Prodigy.* Boole Press, Dublin.

Oersted, H. C. (1812). *Ansicht der chemischen Naturgesetzen durch die neuren Entdeckungen gewonnen.* Realschulbuchhandlung, Berlin. (Reproduced in Ørsted 1920, Vol. 2, pp. 25–169.)

Oersted, H. C. (1821). Betrachtungen ueber den Electromagnetismus. *Journal fuer Chemie und Physik*, **32**, 199–231. (Reproduced in Ørsted 1920, Vol. 2, pp. 223–45.)

Ørsted, H. C. (1820). Experimenta circa effectum conflictus electrici in acum magneticam. Copenhagen. (Reproduced in Ørsted 1920, Vol. 2, pp. 214–18.) German: *Journal für Chemie und Physik* (1820) **29**, 275–9. English: Experiments on the effect of a current of electricity on the magnetic needle. *Annals of Philosophy* (1820) **16**, 273–6.

Ørsted, H. C. (1830). Thermoelectricity. In *The Edinburgh Encyclopaedia*, Vol. 18 (ed. D. Brewster), pp. 573–89. (Reproduced in Ørsted 1920, Vol. 2, pp. 351–98.)

Ørsted, H. C. (1920). *Naturvidenskabelige Skrifter* (ed. Kirstine Meyer, with an introduction on *The Scientific Life and Works of H. C. Ørsted*). Andr. Fred. Host & Søn, Copenhagen.

Öpik, U. and Pryce, M. H. L. (1957). Studies of the Jahn–Teller effect. I. A survey of the static problem. *Proceedings of the Royal Society*, **A238**, 425–47.

Park, D. (1958). Recent advances in physics. *American Journal of Physics*, **26**, 210–34.

Peierls, R. E. (1955). *Quantum Theory of Solids.* Clarendon Press, Oxford.

Peierls, R. E. (1985). *Bird of Passage. Recollections of a Physicist.* Princeton University Press.

Reichenbach, H. (1957). *The Philosophy of Space and Time* (trans. M. Reichenbach and J. Freund). Dover, New York.

Rodrigues, O. (1840). Des lois géometriques qui régissent les déplacements d'un système solide dans l'espace, et la variation des coordonnées provenant de ses déplacements considérés indépendamment des causes qui peuvent les produire. *Journal de Mathématiques Pures et Appliquées*, **5**, 380–440.

Salem, L. (1966). *The Molecular Orbital Theory of Conjugated Systems.* Benjamin, New York.

Shubnikov, A. V. and Kopstik, V. A. (1974). *Symmetry in Science and Art* (trans. G. D. Archard; ed. D. Harker). Plenum Press, New York.

Stauffer, R. C. (1953). Persistent errors regarding Oersted's discovery of electromagnetism. *Isis*, **44**, 307–10.

Stauffer, R. C. (1957). Speculation and experiment in the background of Oersted's discovery of electromagnetism. *Isis*, **48**, 33–50.

Tait, P. G. (1890). *An Elementary Treatise on Quaternions* (3rd edn). Cambridge University Press.

Teller, E. (1972). An historical note. Preface in *The Jahn–Teller Effect in Molecules and Crystals* (by R. Englman), p. v. Wiley-Interscience, London.

Thompson, S. P. (1910). *The Life of William Thomson, Baron Kelvin of Largs.* Macmillan, London.

Weyl, H. (1952). *Symmetry.* Princeton University Press.

Whittaker, Sir Edmund (1951). *A History of the Theories of Aether and Electricity. The Classical Theories* (2nd edn). Nelson, London.

Williams, L. P. (1965). *Michael Faraday.* Chapman & Hall, London.

Wu, C. S., Ambler, E., Hayward, R. W., Hoppes, D. D., and Hudson, R. P. (1957). Experimental test of parity conservation in beta decay. *Physical Review*, **105**, 1413–15.

Index